High Energy Particle Physics Workshop 2017

Journal of Physics: Conference Series Volume 889

Johannesburg, South Africa
1 – 3 February 2017

ISBN: 978-1-5108-4792-7
ISSN: 1742-6588

Printed from e-media with permission by:

Curran Associates, Inc.
57 Morehouse Lane
Red Hook, NY 12571

CURRAN ASSOCIATES INC.
proceedings
.com

Some format issues inherent in the e-media version may also appear in this print version.

Printed by Curran Associates, Inc. (2017)

For permission requests, please contact the Institute of Physics
at the address below.

Institute of Physics
Dirac House, Temple Back
Bristol BS1 6BE UK

Phone: 44 1 17 929 7481
Fax: 44 1 17 920 0979

techtracking@iop.org

Additional copies of this publication are available from:

Curran Associates, Inc.
57 Morehouse Lane
Red Hook, NY 12571 USA
Phone: 845-758-0400
Fax: 845-758-2633
Email: curran@proceedings.com
Web: www.proceedings.com

Table of contents

Volume 889

High Energy Particle Physics Workshop 2017
1–3 February 2017, Johannesburg, South Africa

Accepted papers received: 1 September 2017
Published online: 8 September 2017

Preface

Papers

IOP Conf. Series: Journal of Physics: Conf. Series **889** (2017) 011001 doi:10.1088/1742-6596/889/1/011001

Preface

The motivation for this workshop began with the discovery of the Higgs boson four years ago, and the realisation that many problems remain in particle physics, such as why there is more matter than anti-matter, better determining the still poorly measured parameters of the strong force, explaining possible sources for dark matter, naturalness etc. While the newly discovered Higgs boson seems to be compatible with the Standard Model, current experimental accuracy is far from providing a definitive statement with regards to the nature of this new particle. There is a lot of room for physics beyond the Standard Model to emerge in the exploration of the Higgs boson. Recent measurements in high-energy heavy ion collisions at the LHC have shed light on the complex dynamics that govern high-density quark-gluon interactions. An array of results from the ALICE collaboration has been highlighted in a recent issue of CERN courier. The physics program of high-energy heavy ion collisions promises to further unveil the intricacies of high-density quark-gluon plasma physics.

The great topicality of high energy physics research has also seen a rapid increase in the number of researchers in South Africa pursuing such studies, both experimentally through the ATLAS and ALICE colliders at CERN, and theoretically. Young researchers and graduate students largely populate these research groups, with little experience in presenting their work, and few support structures (to their knowledge) to share experiences with. Whilst many schools and workshops have sought to educate these students on the theories and tools they will need to pursue their research, few have provided them with a platform to present their work. As such, this workshop discussed the various projects being pursued by graduate students and young researchers in South Africa, enabling them to develop networks for future collaboration and discussion.

The workshop took place at the iThemba Laboratories - North facility, in Gauteng, from the 1st to the 3rd of February 2017, where excellent conference facilities with outdoors and indoor tea areas for discussions and interactions were provided, along with a state-of-the-art remote access to the conference venue such that those who were unable to attend the workshop in person could also be present. The laboratory is located on the corner of Jan Smuts Avenue and Empire Road. A morning plenary session, followed 15+10 minute presentations, was the format across our three days. The topics covered being in high-energy theory and phenomenology (heavy ions, pp, ep, ee collisions), ATLAS physics and ALICE physics. The workshop website is http://hep.wits.ac.za/HEPPW2017.php

Alan S. Cornell and Bruce Mellado

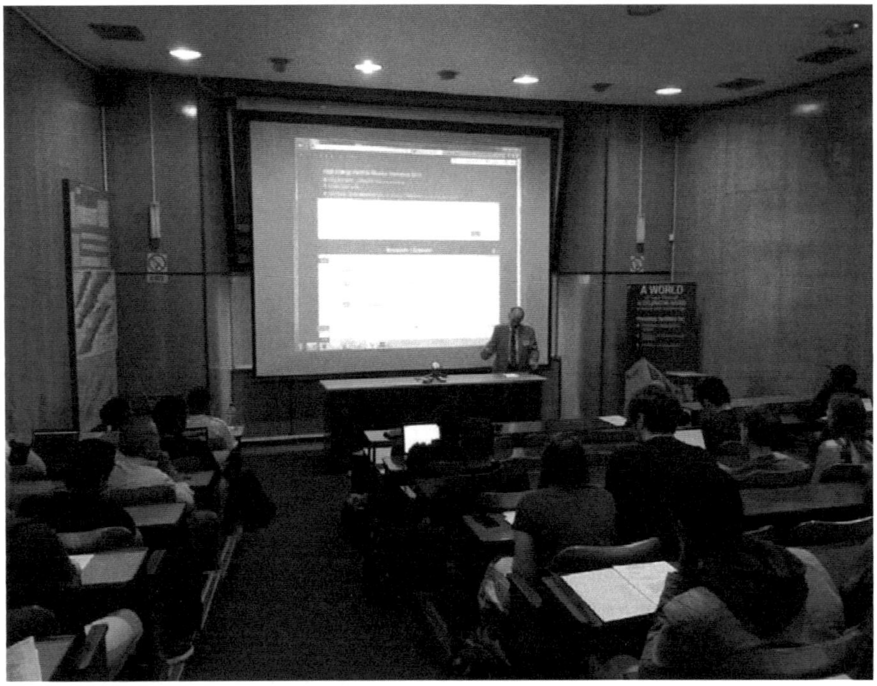

Figure 1. *The opening address delivered by the dean of Science, Prof. Helder Marques.*

Figure 2. *Introductory talk given by the head of SA-CERN, Jean Cleymans.*

Sponsors

- National Institute for Theoretical Physics (NITheP)
- The University of the Witwatersrand
- iThemba Laboratories
- Mandelstam Institute for Theoretical Physics

Organising Committee

Alan S Cornell	NITheP; University of the Witwatersrand - (*co-chair*)
Bruce Mellado	University of the Witwatersrand - (*co-chair*)
Zinhle Buthelezi	iThemba Laboratories
Jean Cleymans	University of Cape Town
Tom Dietel	University of Cape Town
Dino Giovannoni	Rhodes
Will Horowitz	University of Cape Town
Deepak Kar	University of the Witwatersrand
Betty Kibirige	University of Zululand
Xifeng Ruan	University of the Witwatersrand
Azwinndini Muronga	Nelson Mandela Metropolitan University
Elias Sideras-Haddad	Materials Physics Research Institute; University of the Witwatersrand
Sahal Yacoob	University of Cape Town

List of attendees

Amir Abu El Ros	University of the Witwatersrand
Ryan Atkin	University of Cape Town
Geoff Beck	University of the Witwatersrand
Donovan Botlhole	BIUST
Zinhle Buthelezi	iThemba Laboratories
Jean Cleymans	University of Cape Town
Alan Cornell	University of the Witwatersrand
Siegried Fortsch	iThemba Laboratories
Robert Hambrock	University of Cape Town
Gerhard Harmsen	University of the Witwatersrand
Dingane Hlaluku	University of the Witwatersrand
Will Horowitz	University of Cape Town
Abdullah Khalil Hassan Ibrahim	University of Cape Town
Yesenia Jimenez	University of the Witwatersrand

IOP Conf. Series: Journal of Physics: Conf. Series **889** (2017) 011002 doi:10.1088/1742-6596/889/1/011002

Deepak Kar	University of the Witwatersrand
Mohammed Omer Khojali	University of the Witwatersrand
Betty Kibirige	University of Zululand
Jong Soo Kim	IBS - Korea
Andrea Knue	Gottingen
Isobel Kolbe	University of Cape Town
Mukesh Kumar	University of the Witwatersrand
Oscar Kureba	BIUST
Nthabiseng Lekalakala	University of the Witwatersrand
Massimiliano Marchisone	iThemba Laboratories
Helder Marques	University of the Witwatersrand
Senamile Masango	University of Zululand
Thabo Masuku	University of the Witwatersrand
Karabo Mdhluli	University of the Witwatersrand
Joyful Mdluli	University of the Witwatersrand
Bruce Mellado	University of the Witwatersrand
Sibaliso Mhlanga	iThemba Laboratories
Sizwe Mhlongo	University of Zululand
Abdualazem Fadol Mohammed	University of the Witwatersrand
Gaogalalwe Mokgatitswane	BIUST
Tshidiso Molupe	University of the Witwatersrand
Razieh Morad	University of Cape Town
Chuene Mosomane	University of the Witwatersrand
Sinegugu Mthembu	University of Zululand
Skhathi Mthembu	University of the Witwatersrand
Simon Mullins	iThemba Laboratories
Xolane Ngcobo	University of the Witwatersrand
Phuti Ntsoko	University of the Witwatersrand
Tanjona Radonirina Rabemananjara	University of Cape Town
Andriniaina Rasoanaivo	University of Cape Town
Debarati Roy	University of the Witwatersrand
Amal Sarkar	iThemba Laboratories
Justine Tarrant	University of Cape Town
Otsile Tigologo	BIUST
Phuti Tjale	University of the Witwatersrand
Humphry Tlou	University of the Witwatersrand
Kehinde Gbenga Tomiwa	University of the Witwatersrand
Roger van Rensburg	University of the Witwatersrand
Raynette van Tonder	University of Cape Town
Stefan von Buddenbrock	University of the Witwatersrand
Hartmut Winkler	University of Johannesburg

Peer review statement

All papers published in this volume of *Journal of Physics: Conference Series* have been peer reviewed through processes administered by the proceedings Editors. Reviews were conducted by expert referees to the professional and scientific standards expected of a proceedings journal published by IOP Publishing.

Search for a heavy boson in $WW\gamma\gamma$ channel at the Large Hadron Collider in pp at $\sqrt{s} = 13$ TeV and integrated luminosity of 36.5 fb^{-1} with the ATLAS detector

A Fadol[1], B Mellado and X Ruan

School of Physics, University of the Witwatersrand, Johannesburg 2050, South Africa

E-mail: [1]abdualazem.fadol.moohammed@cern.ch

Abstract. In this study, a search for heavy boson H, decaying into the Standard Model Higgs h, and scalar boson S using 36.5 fb^{-1} of proton-proton collision data recorded at $\sqrt{s} = 13$ TeV by the ATLAS detector at the Large Hadron Collider is presented in this note. The investigation is carried out with h decaying into two photons and the semileptonic decays of two W's from S. The analysis procedures are described and the sensitivity of the analysis is presented.

1. Introduction

The discovery of the Standard Model (SM) Higgs boson, h, in 2012, has been announced by the ATLAS and CMS Refs. [1, 2]; in which its spin, CP and charge have been extensively studied [3]. This opened a wide range of research area to investigate whether the discovered Higgs boson is unique in its nature or just one of the spectrum of the SM Higgs. Many theories have been introduced in so-called Beyond the Standard Model of particle physics (BSM), such as the 2HDM [4]. These theories are an extension for the SM, where additional massive Higgs bosons appear. The SM shows the possibility of the self-couplings of the SM Higgs boson. But to observe this production in Run I was impossible as reported in Ref. [5]. This is because of the large background and small signal at low luminosity approximately 20 fb^{-1}. However, physics BSM can significantly enhance the signal yield. For instance, in Ref. [6] a heavy neutral boson has been hypothesized. This heavy boson decays to a pair of scalar bosons, SS, or to SM Higgs and a scalar boson, S, ($gg \to H \to SS, Sh$) as shown in Figure 1 of the resonance case. In this search, the semi-leptonic $WW^*\gamma\gamma$ final state, i.e. with two photons from h, two jets and one charged lepton together with missing energy from S is considered. Hence, the W's bosons decay semileptonically and hadronically as in $gg \to H \to hS \to WW^*\gamma\gamma \to l\nu jj\gamma\gamma$. The work is organized as follows, the description of the data and the Monte Carlo (MC) samples is given in Section 2. The background

IOP Conf. Series: Journal of Physics: Conf. Series **889** (2017) 012001 doi:10.1088/1742-6596/889/1/012001

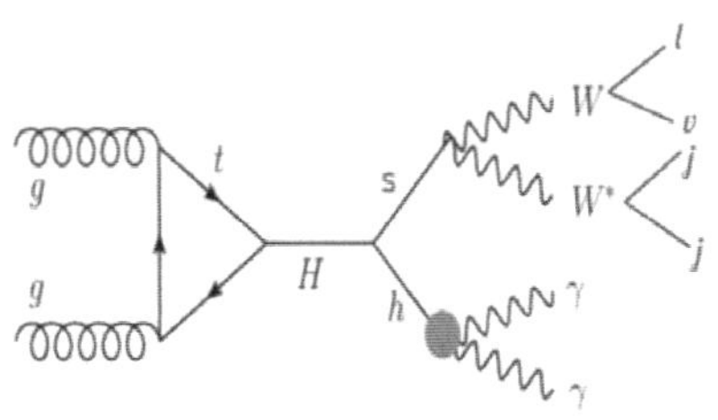

Figure 1. Feynman diagrams for the production mechanism of the heavy boson.

modeling and the spurious signal computation is discussed in Section 4. In Section 5, the signal parameterization is studied. In Section 6 we give preliminary statistical result of the fit on the Asimov data, where the background only hypothesis is used. A conclusion is drawn in Section 7.

2. Data and Monte Carlo samples

The data will be used is the data recorded by ATLAS detector in 2016 at $\sqrt{s} = 13$ TeV with luminosity of 36.5 fb^{-1}. The Monte Carlo samples used to simulate the signal samples are generated by the MadGragh5_aMC@NLO [7]. The resonance mass considered here is between 260-500 GeV, where we assumed that the resonance has small decay width. The background on the SM Higgs is estimated with the following production mechanism: ggh, VBF, Wh, Zh and tth, these samples have been generated with Pythia8 and Powheg by A14 NNPDF2.3LO tunes at center-of-mass energy of 13 TeV. The production cross-section for each sample are 48.52 pb, 3.779 pb, 1.369 pb, 0.8824 pb and 0.5065 pb respectively. Also, events with final states of $l\nu jj\gamma\gamma$ have been simulated with Madgraph5 and Pythia8 for events come from $pp \to l\nu jj\gamma\gamma$ and $pp \to jj\gamma\gamma$.

3. Analysis procedures

The events selections for the two photons are the same as described in Ref. [8]. In addition to the two photons selections, we introduce the following selection:

- The number of jets to be at least two central jets which are required to have $p_T^{jets} > 25$ GeV and $|\eta| < 2.5$.
- An event will be rejected if any b-tagged jet is found. The b-tagger is MV2c10 with a b-tagging efficiency of 70%.
- The number of leptons in the final state should be at least one lepton and it could be either a muon or an electron.
- The tight mass window is defined by $(125.09 - 2 \times 1.7) \leq m_{\gamma\gamma} \leq (125.09 + 2 \times 1.7)$.

The selection efficiencies for the signal samples as a function of the mass of H are shown in Table 1. Where the selection efficiency at the mass, $m_{\gamma\gamma}$, cut is between 35.2-41.5%.

Table 1. The cut efficiency for the signal samples.

Cuts \ Samples	S and h	Resonance Mass			
		260 GeV	300 GeV	400 GeV	500 GeV
All Events	100.0%	100.0%	100.0%	100.0%	100.0%
Duplicate	100.0%	100.0%	100.0%	100.0%	100.0%
GRL	100.0%	100.0%	100.0%	100.0%	100.0%
Pass Trigger	71.5%	67.3%	67.7%	69.7%	72.6%
Detector Quality	71.5%	67.3%	67.7%	69.7%	72.6%
has PV	71.5%	67.3%	67.7%	69.7%	72.6%
2 Loose Photons	59.0%	56.2%	55.7%	57.2%	59.7%
Trig Match	58.7%	55.9%	55.5%	56.9%	59.4%
Tight ID	49.1%	45.8%	45.3%	47.2%	50.0%
Isolation	44.3%	38.4%	38.8%	42.3%	45.6%
Rel. Pt cuts	40.8%	35.8%	35.1%	38.4%	42.4%
$105 < m_{\gamma\gamma} < 160$	39.9%	35.2%	34.5%	37.6%	41.5%
At least 2 central jets	31.1%	19.8%	22.3%	28.2%	33.2%
B-veto	29.1%	18.8%	21.1%	26.5%	31.0%
At least 1 lepton	5.9%	3.5%	3.9%	5.4%	6.3%
Tight mass window	5.10%	2.94%	3.36%	4.63%	5.52%

Table 2. The selection efficiency for the background samples.

	ggH	VBF	WH	ZH	ttH
All Events	100.0%	100.0%	100.0%	100.0%	100.0%
Duplicate	100.0%	100.0%	100.0%	100.0%	100.0%
GRL	100.0%	100.0%	100.0%	100.0%	100.0%
Pass Trigger	61.6%	63.3%	58.4%	57.9%	74.6%
Detector Quality	61.6%	63.3%	58.4%	57.9%	74.6%
has PV	61.6%	63.3%	58.4%	57.9%	74.6%
2 Loose Photons	52.6%	54.0%	47.0%	47.7%	61.2%
Trig Match	52.5%	53.9%	46.9%	47.6%	60.8%
Tight ID	45.9%	46.9%	40.5%	41.1%	50.8%
Isolation	41.1%	42.4%	35.8%	36.3%	42.0%
Rel. Pt cuts	38.2%	38.5%	32.8%	33.2%	38.4%
$105 < m_{\gamma\gamma} < 160$	38.0%	38.1%	32.4%	32.8%	37.7%
At least 2 central jets	5.84%	10.74%	15.81%	16.80%	37.08%
B-veto	5.57%	10.18%	15.06%	13.75%	6.70%
At least 1 lepton	0.00390%	0.01108%	0.57536%	0.38308%	2.06%
Tight mass window	0.00363%	0.01005%	0.52710%	0.35394%	1.92%

After additional cuts, the selection efficiencies for the signal is between 2.94-5.52% for the resonance mass from 260 GeV to 500 GeV. Also the selection efficiencies for the SM Higgs background is listed in Table 2. In this table, most of the contribution comes from the tth, and this is due to the top quark, which decays to jets and leptons.

4. Spurious signal

The method of the spurious signal is to help us decide which probability density function (PDF) will be suitable for our study. In addition, It allows us to estimate the signal systematics due to the mis-modeling of the background

Table 3. Spurious signal computation and the selection result of the PDF for the background.

Function	S	$Max(S/\Delta S)$	Result	Selection
Exponential	-0.50	-13.2%	Pass	Selected
Dijet	0.23	6.0 %	Pass	-
ExpPoly2	-0.03	-0.7 %	Pass	-
Bern3	-0.03	-0.6 %	Pass	-

In order to perform the spurious signal study we have to prepare an input as a histogram that stored in a ROOT file. The spurious signal here will be computed only on the range of the Higgs signal. The model uses for the fitting is $S+B$, which has the composite PDF form:

$$N_S \times PDF_S + N_B \times PDF_B . (1)$$

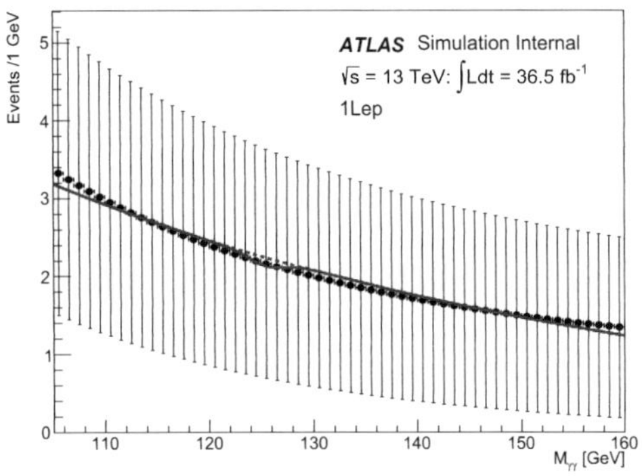

Figure 2. The $S+B$ fit for the selected exponential PDF.

Where N_S, PDF_S, N_B and PDF_B are number of signal, PDF of the signal, number of background and PDF of the background, respectively. The PDF of the signal is already stored in the workspace and the PDF of the background is to be determine by this study. The selection of the background PDF is as follows:

- A fit of $(S+B)$ will be implemented first on the input MC for background, so that we can get the number of the signal, N_S, for all the selected background PDF's.
- The maximum fitted signal will be recorded.
- Hence, we apply selection as to whether to accept or reject a certain PDF. This achieved by the maximum signal over the error to be less than 20%, $Max(S/\Delta S) <$ 20%.
- The PDF is chosen if it passed the condition above. However, if all the selected function are passed the selection criteria, then the PDF with less parameters will be selected by the tool.
- In the configuration file the integrated luminosity of the input dataset has to be consistent with the luminosity of the computed yield.
- The lower and upper end of the window to maximize over to compute the spurious signal is (Higgs mass$\pm$ 8) GeV.
- First and second order Exponential, Bernstein and Dijet are the PDF to be tested.
- The scan range has been set to be within the range of (115-135) GeV.
- The scan is performed by taking 0.5 as interval between the scan points.

The result of the PDF selection depending upon the criteria discussed above for the background is shown in Table 3.

5. Signal parametrization

We used the double-sided Crystal Ball to model the shape of the signal and to get the parameters of the best fit.

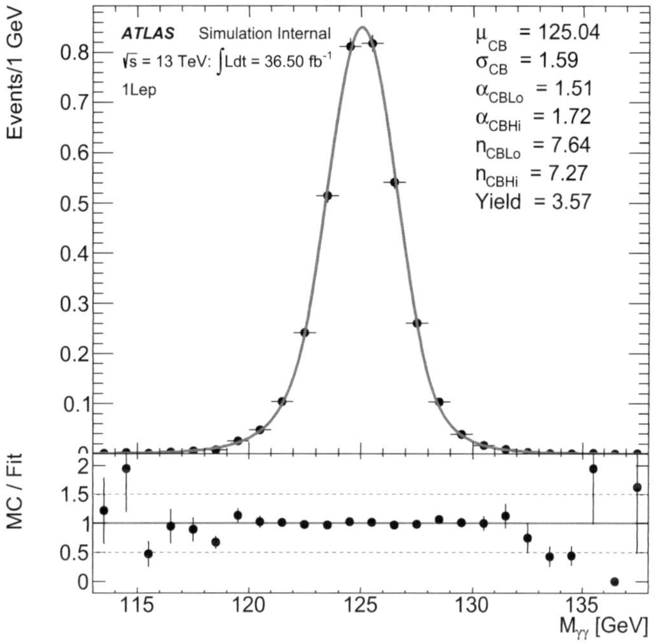

Figure 3. Events pass the selections defined in the signal region.

The single signal and multiple signal fit is performed around the Higgs mass signal, $m_H = 125$ GeV, for the one lepton category, as shown in Figure 3.

6. Statistical results

An Asimov test is performed to test the expected significance as shown in Figure 4, where we found that the expected significance to be $1.43\,\sigma$. The Asimov data is generated as "$Background + 1 \times SM\ Higgs + a\ Heavy\ Boson\ Signal$".

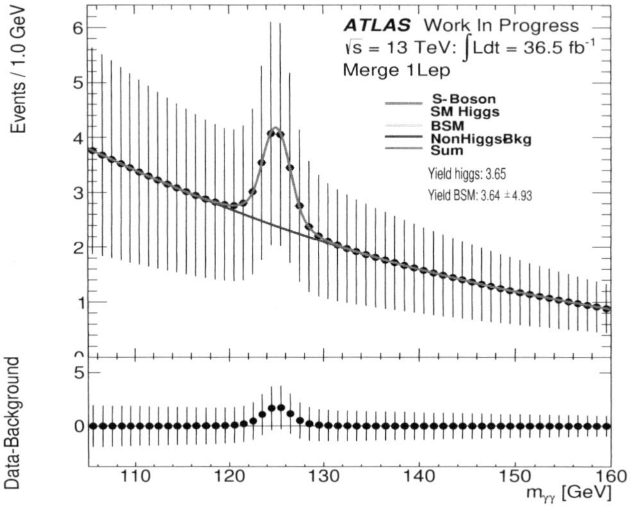

Figure 4. A $(S + B)$ fit for the Asimov data. The Higgs yield is 3.65 and the BSM yield is 3.64±4.93.

7. Conclusion

This search shows preliminary study for the presence of a heavy boson, H, in $WW\gamma\gamma$ channel. Because the data still blinded we use the Asimov data. The analysis is based on the data that is collected at the ATLAS detector at 13 TeV and luminosity of 36.5 fb^{-1}. This is to improve the analysis and test the expected significance, which is found to be $1.43\,\sigma$ with exclusion limit of 3.02 pb.

References

[1] ATLAS Collaboration, Observation of a new particle in the search for the Standard Model Higgs boson with the ATLAS detector at the LHC, Phys. Lett. B **716**, 1 (2012) [arXiv:1207.7214].
[2] CMS Collaboration, Observation of a new boson at a mass of 125 GeV with the CMS experiment at the LHC, Phys. Lett. B **716**, 30 (2012) [arXiv:1207.7235].
[3] ATLAS Collaboration, Measurements of the properties of the Higgs boson using the ATLAS detector, EPJ Web Conf. **126**, 04038 (2016).
[4] G. C. Branco, P. M. Ferreira, L. Lavoura, M. N. Rebelo, M. Sher and J. P. Silva, Theory and phenomenology of two-Higgs-doublet models, Phys. Rept. **516**, 1 (2012) [arXiv:1106.0034].
[5] ATLAS Collaboration, Searches for Higgs boson pair production in the $hh \to bb\tau\tau, \gamma\gamma WW^*, \gamma\gamma bb, bbbb$ channels with the ATLAS detector, Phys. Rev. D **92**, 092004 (2015) [arXiv:1509.04670].
[6] S. von Buddenbrock *et al.*, Phenomenological signatures of additional scalar bosons at the LHC, Eur. Phys. J. C **76**, no. 10, 580 (2016) [arXiv:1606.01674].
[7] J. Alwall, M. Herquet, F. Maltoni, O. Mattelaer and T. Stelzer, MadGraph 5 : Going Beyond, JHEP **1106** (2011) 128 [arXiv:1106.0522].
[8] ATLAS Collaboration, Measurement of Higgs boson production in the diphoton decay channel in pp collisions at center-of-mass energies of 7 and 8 TeV with the ATLAS detector, Phys. Rev. D **90**, no. 11, 112015 (2014) [arXiv:1408.7084].

IOP Conf. Series: Journal of Physics: Conf. Series **889** (2017) 012002 doi:10.1088/1742-6596/889/1/012002

Initial State Factorization and the Kinoshita-Lee-Nauenberg Theorem

Abdullah Khalil and W. A. Horowitz

Department of Physics, University of Cape Town, Rondebosch 7701, South Africa

E-mail: abdullah@aims.ac.za

Abstract. We carry out for the first time a fully correct, self-consistent application of the Kinoshita-Lee-Nauenberg theorem in a next-to-leading order scattering calculation. We improve on previous works by including all initial and final state soft radiative processes, including absorption and an infinite sum of partially disconnected amplitudes. We show in general that the sum over the initial state radiation factorizes. We apply our approach to the Rutherford scattering cross section at next-to-leading order correction.

1. Introduction

The infrared (IR) divergences in purely massless theories come in two flavors: *soft*, due to the massless nature of the radiation, and *collinear*, which comes from treating the radiating particle as massless. The Bloch-Nordsieck (BN) theorem [1, 2] is known as the first theoretical response towards solving the IR problem by summing over the final degenerate states including the emission of soft photons within the energy window Δ. The BN theorem gave an exact cancellation of the soft IR divergences leaving behind some collinear singularities. The non-cancelling collinear singularities have been studied later separately by Kinoshita [3], Lee and Nauenberg [4] (KLN) where they came up with a more general theorem which states that one should sum over both initial and final indistinguishable degenerate states in order to get rid of all the collinear divergences. Lee and Nauenberg (LN) included an example demonstrating the IR finite cross section for elastic Rutherford scattering computed at next-to-leading order(NLO) when one integrates over a degenerate final-state energy window and degenerate initial and final state angular windows. However, for consistency, one should also sum over the initial degenerate soft photon states which spoil the IR cancellation.

Many attempts has been made to lead the non-cancelling IR divergences issue from the initial state soft absorption by invoking the role of the disconnected diagrams, where the disconnected amplitudes can interfere with fully connected amplitudes in which there are both emission and absorption of a soft photon, forming a connected cut diagram [5–9]. However, the former subsequent works summed over some, but not all, possible degenerate initial and final states, a full discussion on the incompleteness of their calculations can be found in Ref. [7]. Ito [10] and Akhoury, Sotiropoulos, and Zakharov [11] (IASZ) included all possible initial and final degenerate states with an infinite number of disconnected soft photons forming a divergent infinite series at NLO. They proved that by rearranging the series through formal manipulations, it will be sufficient to some over a finite number of diagrams, but, as was pointed out in [7], the IASZ rearrangement yields a NLO result that is exactly 0 at tree level.

We will give an example where an electron is scattered off of a point charge by including the NLO diagrams excluding the vacuum polarization and the box correction. We show that the resolution is to use the IASZ approach. In contrary to the IASZ rearrangement, we provide a different rearrangement under a control of a convergent factor to prove the uniqueness of our result, we note that the disconnected part remains factorized.

2. NLO Rutherford

Consider an electron with four-momentum $p = (E, \vec{p})$ scatters off a static point charge into a state with final four-momentum $p' = (E, \vec{p}')$, the momentum transfer is then $q \equiv p' - p$. We use the photon mass m_γ and keep the electron mass m_e non-zero to regulate the soft and collinear IR divergences respectively. The application of the BN theorem requires including the tree level Fig. 1 (a) and NLO vertex correction (b) as well as the final state soft bremsstrahlung process (c) and (d). Summing these processes yields

$$d\sigma_{BN} = d\sigma_0 \left\{ 1 + \frac{\alpha_e}{\pi^2} \left[\log\left(\frac{E^2}{\Delta^2}\right) \left(1 - \log\left(\frac{-q^2}{m_e^2}\right)\right) + \frac{3}{2} \log\left(\frac{-q^2}{m_e^2}\right) + \mathcal{O}(1) \right] \right\}, \tag{1}$$

which is free of soft IR divergences. We now use the KLN theorem to remove the potentially large logs, $\log(-q^2/m_e^2)$, in the high energy limit $(-q^2 \gg m^2)$ by summing over the indistinguishable hard collinear final emission Fig. 1 (d) and initial absorption (e) processes. We note that, the indistinguishably fro such a process is determined by the angular resolution (δ). One can easily see that the LN result can be obtained by replacing the m_e^2 by $\delta^2 E^2$ [4], and the result is free of any IR divergences. Here, we point out to the fact that we used a combination between the BN and KLN theorems for such a cancellation, which is clearly inconsistent [7], and to stay in the spirit of the KLN theorem one should also add the absorption of a soft photon as a degenerate initial state which devastate the finiteness of the cross section at NLO.

It is also shown that including the contributions from the interference of the disconnected diagram Fig. 1 (g) with the absorption-emission diagrams Fig. 1 (h) is not sufficient for the cancellation, even adding the contributions from the absorption or emission with one disconnected photon keeps the cross section IR divergent and raises the question of adding even more disconnected photons to the former processes, while the most important question now: is there a full systematic method to apply the KLN theorem properly? We show here that the resolution is to include all possible contributions from degenerate initial and final states, including the contributions from the disconnected diagrams with an arbitrary number of disconnected soft photons. Following [10, 11], we consider a general form of our process with m incoming soft photons and n outgoing soft photons, each process is described by the transition probability $\mathcal{P}_{mn}$ corresponding to an amplitude $\mathcal{M}_{mn}$. The total Lee-Nauenberg probability will be

$$\mathcal{P} = \sum_{m,n=0}^{\infty} \mathcal{P}_{mn} = \sum_{m,n=0}^{\infty} \frac{1}{m!} \frac{1}{n!} \sum_{i,f} |\mathcal{M}_{mn}|^2, \tag{2}$$

It is shown in [10, 11] that any cut diagram from $\mathcal{P}_{mn}$ at NLO correction can be constructed from a four essential probabilities: $\mathcal{P}_{00}$ which is the cut diagram with no real emission or absorption of soft photons, this may include the leading term, the vertex correction, the vacuum polarization, etc.; $\mathcal{P}_{10}$ and $\mathcal{P}_{01}$ which includes all cut diagrams with one soft photon in the initial or the final states respectively; $\mathcal{P}_{11}$ which includes all cut diagrams from the interference between the disconnected diagram Fig. 1 (g) and the absorption-emission diagrams like Fig. 1 (h). The fully connected cut diagrams are given by any of the previous basic probabilities while the partially connected ones are these probabilities multiplied by a number of δ functions

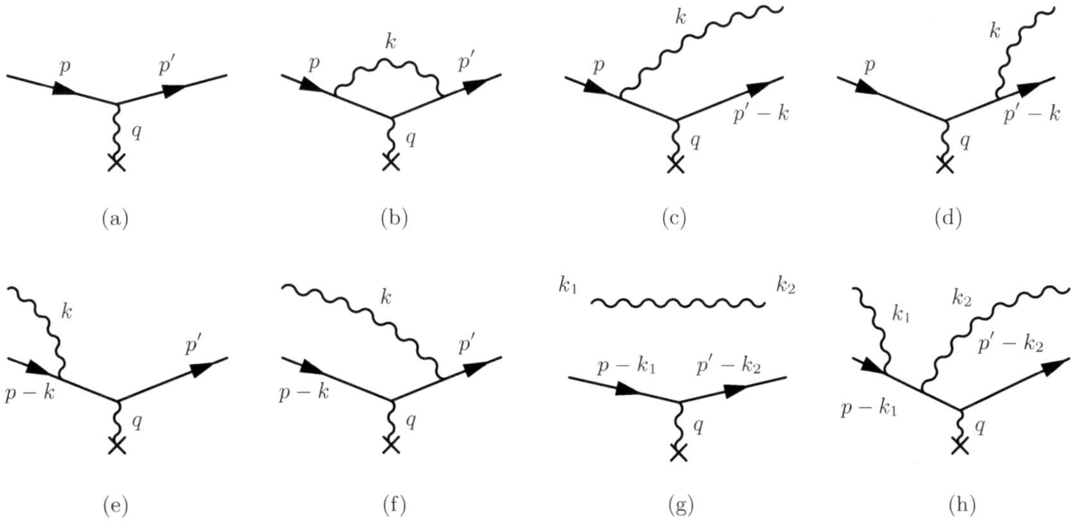

Figure 1. The (a) leading order (LO) and (b) next-to-leading order (NLO) vertex contributions to the $1 \to 1$ process in Rutherford scattering. The (c) and (d) potentially degenerate final state radiative processes. The (e) and (f) potentially degenerate initial state radiative processes. A disconnected photon amplitude such as (g) might interfere with an amplitude with a photon absorption and emission process such as (h).

according to the number of disconnected photons, so we can construct $\mathcal{P}_{mn}$ by splitting each cut diagram up into connected and disconnected parts. We define the disconnected part by the function $\mathcal{D}(m-i, n-j)$ that describes the number of $m-i$ incoming and $n-j$ outgoing soft photons that can be joined together and become disconnected from the electron line. By definition $\mathcal{D}(0,0) = 1$ and $\mathcal{D}(a,b) = 0$ for $a \neq b$. One may show that [10, 11] the transition probability for the general process at NLO is given by

$$\mathcal{P}_{mn} = \frac{\mathcal{D}(m,n)}{m!\,n!}\mathcal{P}_{00} + \sum_{i=0} \frac{\mathcal{D}(m-i, n-i-1)}{(m-i)!\,(n-i-1)!}\mathcal{P}_{01} + \sum_{i=0} \frac{\mathcal{D}(m-i-1, n-i)}{(m-i-1)!\,(n-i)!}\mathcal{P}_{10} + \sum_{i=0} \frac{\mathcal{D}(m-i-1, n-i-1)}{(m-i-1)!\,(n-i-1)!}\widetilde{\mathcal{P}}_{11}. \quad (3)$$

IASZ were able to rearrange the full series to give an IR finite value for $\mathcal{P}$. However, as it is shown in [7] that their argument is not safe because the probability $\mathcal{P}$ tends to zero once we send it back to the tree level. One may show that the series can be rearranged in different ways. However, we are looking for a rearrangement with special features: to be IR safe and to keep the tree level contribution finite. We give here the rearrangement that satisfies these criteria and rigorously prove that our result is unique. In order to give such a proof, we manipulate Eq. (3) in a controlled way by introducing a convergence factor that becomes small for large i: we take

$$\mathcal{D}(m-i, n-j) \;\; \to \;\; \mathcal{D}(m-i, n-j)e^{-(i+j)/\Lambda} \quad (4)$$

with $\Lambda \gg 1$. The convergent factor Λ allows us to sum up to finite number of disconnected soft photons N. Since $\mathcal{D}(m,n) = 0$ for $m \neq n$, we may simplify our manipulations by replacing the double sum over m and n with a single sum over n. With the above convergence factor, we

IOP Conf. Series: Journal of Physics: Conf. Series **889** (2017) 012002 doi:10.1088/1742-6596/889/1/012002

are guaranteed that $\mathcal{P} = \lim_{\Lambda\to\infty} \lim_{N\to\infty} \mathcal{P}_N(\Lambda)$ converges. Now, we rearrange the partial sum $\mathcal{P}_N(\Lambda)$ up to N soft photons to find

$$\mathcal{P}_N(\Lambda) = \sum_{n=0}^{N} \frac{\mathcal{D}(n,n)}{(n!)^2} \left[\mathcal{P}_{00} + e^{-\frac{1}{\Lambda}}\mathcal{P}_{01}\right] + \sum_{n=1}^{N}\sum_{i=1}^{n} \frac{\mathcal{D}(n-i,n-i)}{[(n-i)!]^2} \left[e^{-\frac{2i+1}{\Lambda}}\mathcal{P}_{01} + e^{-\frac{2i-1}{\Lambda}}\mathcal{P}_{10} + e^{-\frac{2i}{\Lambda}}\widetilde{\mathcal{P}}_{11}\right], \quad (5)$$

One can easily show, using the same convergent factor, that the IASZ result is very different from Eq. (5) after swapping the limit in the full sum. So we check the possibility of switching the limits in order to prove the uniqueness of our result. We exploit the Monotone Convergence Theorem to justify that our result converges to the same value under the interchanging limits procedure by proving that the partial sum $\mathcal{P}_N(\Lambda)$ 1) monotonically increase in N for each Λ and 2) monotonically increase in Λ for each N [12]. We use the fact that $\mathcal{P}_{01} + \mathcal{P}_{10} = -\widetilde{\mathcal{P}}_{11}$ [6] to simplify Eq. (5). Then we have

$$\mathcal{P}_N(\Lambda) = \sum_{n=0}^{N} \frac{\mathcal{D}(n,n)}{(n!)^2} \left[\mathcal{P}_{00} + e^{-\frac{1}{\Lambda}}\mathcal{P}_{01}\right] + \sum_{n=1}^{N}\sum_{i=1}^{n} \frac{\mathcal{D}(n-i,n-i)}{[(n-i)!]^2}\, 2\mathcal{P}_{01}\, e^{-\frac{2i}{\Lambda}} \left[\cosh\left(\frac{1}{\Lambda}\right) - 1\right]. \quad (6)$$

Since $\mathcal{P}_{00}$, $\mathcal{P}_{01}$, Λ, and $\mathcal{D}(n,n)$ are all strictly positive, Eq. (6) clearly increases monotonically in N for fixed Λ. To show that $\mathcal{P}_N(\Lambda)$ increases monotonically in Λ for fixed N, we take the derivative with respect to Λ:

$$\frac{d\mathcal{P}_N(\Lambda)}{d\Lambda} = \sum_{n=0}^{N} \frac{\mathcal{D}(n,n)}{(n!)^2} \left[\frac{1}{\Lambda^2}\, e^{-\frac{1}{\Lambda}}\, \mathcal{P}_{01}\right] + \mathcal{O}\left(\frac{1}{\Lambda^3}\right). \quad (7)$$

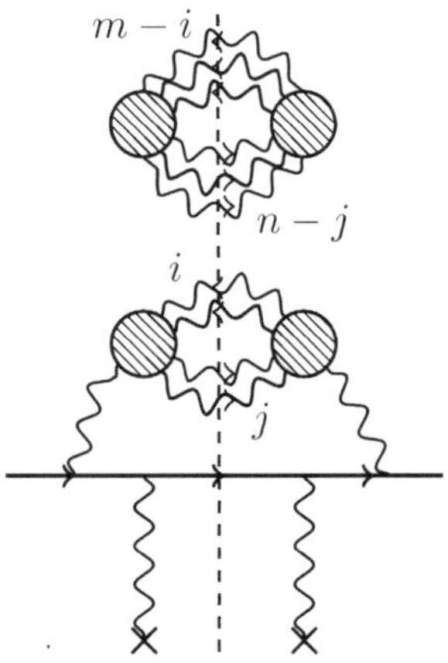

Figure 2. A generic cut diagram with m incoming and n outgoing soft photons that is partially disconnected. (Note that the blobs include the possibility of connecting incoming photon lines from the left with conjugate incoming photon lines from the right.)

Although one finds that the higher order in $1/\Lambda$ correction term is negative, for any N we can find a Λ large enough such that the first term, which is strictly positive, dominates. We have thus proved that we may exchange limits for our rearranged formula Eq. (6), and we may evaluate the $\Lambda \to \infty$ limit first, yielding our main result:

$$\mathcal{P} = (\mathcal{P}_{00} + \mathcal{P}_{01}) \sum_{n=0}^{\infty} \frac{\mathcal{D}(n,n)}{(n!)^2}. \quad (8)$$

We find that all the soft initial state physics of the infinite number of undetectably soft photons completely factorizes. When the cross section is computed, one simply divides out by this unobserved infinity. We have thus rendered all soft and collinear IR divergences harmless.

3. Conclusions

A self-consistent application of the KLN theorem requires a sum over all degenerate initial and final states to arrive at an IR safe cross section. We checked the IR cancellation to the Rutherford scattering cross section at NLO correction by including all possible diagrams with an arbitrary number of disconnected soft photons which forms a divergent infinite series. Introducing a convergent factor helped us to manipulate the formally divergent series under control, which helps us to apply the Monotone Convergence Theorem to prove that the given rearrangement is unique through interchanging the limits in the full sum. Our rearrangement shows that we only need to sum up t a finite number of diagrams where the disconnected part remains factorized, and more important that the sum of these diagrams is IR safe and leads to the tree level contribution when we take it back to the LO. We, therefore, arrived at the extremely nontrivial result that for NLO Rutherford scattering the summation over *all* indistinguishable initial and final states is equivalent to the summation over *only* the initial hard collinear and final soft, hard collinear, and soft and collinear degenerate states.

Acknowledgments

We thank the South African National Research Foundation and the SA-CERN consortium for their support. AK also thanks the African Institute for Mathematical Sciences (AIMS) and the University of Cape Town for their support.

References

[1] F. Bloch and A. Nordsieck, "Note on the Radiation Field of the electron," *Phys. Rev.*, vol. 52, pp. 54–59, 1937.

[2] D. R. Yennie, S. C. Frautschi, and H. Suura, "The infrared divergence phenomena and high-energy processes," *Annals Phys.*, vol. 13, pp. 379–452, 1961.

[3] T. Kinoshita, "Mass singularities of Feynman amplitudes," *J. Math. Phys.*, vol. 3, pp. 650–677, 1962.

[4] T. D. Lee and M. Nauenberg, "Degenerate Systems and Mass Singularities," *Phys. Rev.*, vol. 133, B1549–B1562, 1964, [,25(1964)].

[5] C. De Calan and G. Valent, "Infra-red divergence and incident photons," *Nucl. Phys.*, vol. B42, pp. 268–280, 1972.

[6] T. Muta and C. A. Nelson, "Role of Quark - Gluon Degenerate States in Perturbative QCD," *Phys. Rev.*, vol. D25, p. 2222, 1982.

[7] M. Lavelle and D. McMullan, "Collinearity, convergence and cancelling infrared divergences," *JHEP*, vol. 03, p. 026, 2006. arXiv: `hep-ph/0511314 [hep-ph]`.

[8] B. Mirza and M. Zarei, "Cancellation of soft and collinear divergences in noncommutative QED," *Phys. Rev.*, vol. D74, p. 065 019, 2006. arXiv: `hep-th/0609181 [hep-th]`.

[9] M. Lavelle, D. McMullan, and T. Steele, "Soft Collinear Degeneracies in an Asymptotically Free Theory," *Adv. High Energy Phys.*, vol. 2012, p. 379 736, 2012. arXiv: `1008.3949 [hep-ph]`.

[10] I. Ito, "INFRARED DIVERGENCES IN QCD AND KINOSHITA-LEE-NAUENBERG THEOREM," *Prog. Theor. Phys.*, vol. 67, p. 1216, 1982.

[11] R. Akhoury, M. G. Sotiropoulos, and V. I. Zakharov, "The KLN theorem and soft radiation in gauge theories: Abelian case," *Phys. Rev.*, vol. D56, pp. 377–387, 1997. arXiv: `hep-ph/9702270 [hep-ph]`.

[12] A. W. Knapp, *Basic real analysis*. Springer Science & Business Media, 2005.

IOP Conf. Series: Journal of Physics: Conf. Series **889** (2017) 012003 doi:10.1088/1742-6596/889/1/012003

Correction to the Probability Distribution of Induced Gluon

Andriniaina N. Rasoanaivo & W.A. Horowitz

Department of Physics, University of Cape Town, Private Bag X3, Rondebosch 7701, South Africa.

E-mail: andriniaina@aims.ac.za

February 2017

Abstract. We present a novel technique for the calculation of the probability for emission of gluon radiated from a high-p_T quark through a medium (QGP). Our work is an extension of the maximal helicity violating (MHV) method to compute the non-abelian correction for 2 gluon emissions.

1. Introduction

It is believed that shortly after the big bang the universe was filled by mixture of quarks and gluons which are the fundamental building blocks of matters. The mixture is known as the quark gluon plasma (QGP) where quarks and gluons are no longer confined. And as the universe expanded the temperature decreases and the interaction increases, then quarks and gluons start to combine into baryons and mesons which will become atoms then molecules and eventually stars and galaxies will emerge.

In heavy ion collision, at RHIC and LHC, experiments have been conducted in order to reproduce the condition at early universe where QGP can be probed [1]. In these experiments heavy ions like Au or Pb are accelerated and collide them at very high speed, so every single proton and neutron will melt into free quark and gluon. One of the challenge in the experimental side is to detect any formation of QGP from the collision, since QGP has a very short lifetime and cannot be directly observed. This is one of the reasons why we are interested on "jets", since they carry information on any medium produced in heavy ion collisions.

Jets are the manifestation of high-p_T partons produced from nucleus-nucleus collision. These partons experience multiple interaction inside the collision region prior to fragmentation and hadronisation, especially in presence of medium, jet looses energy through scattering and mainly gluon radiation [2, 3, 4]

$$\text{Jet}(E) \longrightarrow \text{Jet}(E - \Delta E) + \text{Radiation}(\Delta E). \tag{1}$$

Practically, we cannot detect those radiations since they are either emitted at low energy compare to the resolution of the detectors (*soft*), or they are emitted at a very small angle and they cannot be resolved from the parent jet (*collinear*). However in the theoretical side, the soft-collinear behaviour of these radiations provide the necessary tools to compute the probability

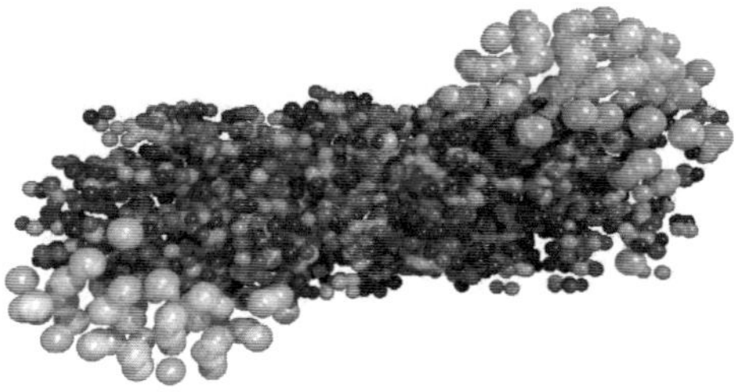

Figure 1: Heavy ion collisions to probe quark gluon plasma.

distribution for such emission. But due to the complexity of QCD (quantum chromodynamic) the probability of gluon radiation was approximated to be poisson, in another word the gluon interferences was neglected and the individual emission are independent from each other. Then the main goal of this work is to perform a full QCD calculation and find the non-abelian correction that comes from gluon interferences to the poissonian approximation,

$$dP_{\text{improved}} = dP_{\text{poisson}} + dP_{\text{correction}}. \tag{2}$$

In this proceeding, we consider a light quark passing through a medium tickled by a gluon and compute the probability distribution two gluon emission. In the next section, we are going to give a short introduction to the maximal helicity violating (MHV) techniques that we are going to use. Then in section3 we are going to compute the two gluon emission in the poisson approximation, and last in section4 we present the non-abelian correction for the two gluon emissions.

2. Maximal Helicity Violating

This section is short introduction for the maximal helicity violating techniques in order to compute scattering amplitudes [7, 8, 9]. As physicist, it is very natural for us to use Feynman diagrams techniques to compute amplitudes, but we know that the computations get complicated as the number of particle increases. For example in the case of photon to quark anti-quark production, we can see in the table below how the number of digram increases exponentially with the number of gluon corrections

# Gluon	1	2	3	⋯	n
# Diagram	2	8	48	⋯	$2^n n!$

But it turns out, the analytical expression of the amplitudes are extremely simple using the spinor helicity formalism and with a simple change of variables especially when the conservation of helicity is maximally violated, i.e. $\Delta h = |h_i - h_f|$ is maximal. The spinor helicity formalism is one of the main ingredient of the MHV techniques, and in that formalism instead of the four vector representation the momenta will be represented with a 2 by 2 matrices given by $P^{a\dot{a}} = (\sigma_\mu)^{a\dot{a}} p^\mu$. And the invariant mass of the individual particle will be the determinant of the respective momenta in that representation. For massless particles, since de invariant mass is zero then the momenta can be parametrized as follow

$$P^{a\dot{a}} = (\sigma_\mu)^{a\dot{a}} p^\mu = \lambda^a \widetilde{\lambda}^{\dot{a}}, \tag{3}$$

where λ and $\widetilde{\lambda}$ are called the spinor helicity variables and transform respectively as left and right handed spinors. Out of those new variables, we can define two different Lorentz invariant

IOP Conf. Series: Journal of Physics: Conf. Series **889** (2017) 012003 doi:10.1088/1742-6596/889/1/012003

products which are both antisymmetric

$$\langle 12\rangle = \epsilon_{ab}\lambda_1^a\lambda_2^b \quad \text{and} \quad [12] = \epsilon_{\dot{a}\dot{b}}\widetilde{\lambda}_1^{\dot{a}}\widetilde{\lambda}_2^{\dot{b}}, \tag{4}$$

those products, angle and square brackets, can be related to the four vector scalar product by $2p_1.p_2 = \langle 12\rangle\,[12]$. The MHV amplitude corresponding to a tickled quark that radiate n gluon is given by the partial amplitude below,

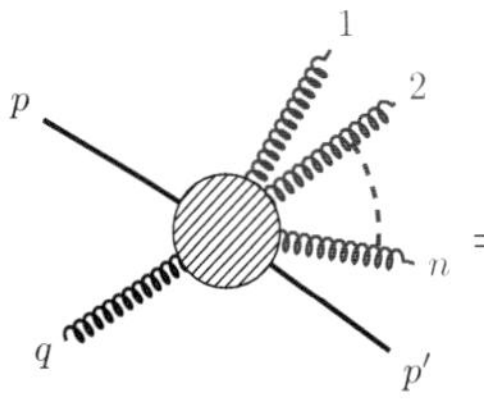

$$= g^{n+1}\frac{\langle pq\rangle^3\langle qp'\rangle}{\langle pp'\rangle\langle p'q\rangle\langle q1\rangle\langle 12\rangle\cdots\langle np\rangle} \tag{5}$$

The full amplitudes can be computed by summing the permutation of gluon indices weighted by the color string basis $(T_{a_1}T_{a_2}\cdots T_{a_n})$.

3. Two gluon emissions in the poisson approximation

In the poisson approximation it is assumed that gluon emissions are independent then the distribution of multiple gluon emission is given by the product of individual emissions. Technically, independent emission is satisfied when the emission angles of the gluon emitted are strongly ordered also known by *angular ordering*, $\theta_1 \gg \theta_2 \gg \cdots \gg \theta_n$ where θ_i is the emission angle of the i-th gluon. To begin let us recall the soft-collinear factorization where the information on the momentum distribution is factorized from the born amplitude as below,

$$\mathcal{M}_{n\text{-gluon}} = \mathcal{M}_{\text{born}} \times J_n(\{k_i\}), \tag{6}$$

where $\mathcal{M}_{\text{born}}$ is the born amplitude and J_n is the eikonal factor. Now using this soft-collinear factorization using the MHV techniques, we can compute the single gluon distribution $dP^{(1)}$ to be the square of the eikonal factor J_1 and we obtain

$$\frac{dP^{(1)}}{d\Phi_1} = \frac{C_A}{2}\frac{p.q}{2p.k_1 q.k_1}, \tag{7}$$

where $d\Phi_1$ is the differential phase space of single emission of momentum k_1, and C_A is the color factor and equal to 3 for $SU(3)$. And if have independent emission the 2 gluon distribution is given by the following expression,

$$\frac{dP^{(2)}}{d\Phi_{1,2}} = \frac{1}{2}\left(\frac{C_A}{2}\right)^2 \prod_{I=1,2}\frac{p.q}{2p.k_I q.k_I}, \tag{8}$$

$d\Phi_{1,2}$ is the differential phase space of two gluons of momenta k_1 and k_2. In the next section we will use MHV to fully compute the two gluon emission distribution and show that the poisson approximation is only valid in some region of the phase space.

4. Non-abelian correction for the two gluon emissions

To compute the full amplitude in the MHV, we need perform a sum over the permutation of gluon indices weighted by the color string bases in which we apply the factorization. After

IOP Conf. Series: Journal of Physics: Conf. Series **889** (2017) 012003 doi:10.1088/1742-6596/889/1/012003

computation, the eikonal factor J_2 can be written in two terms which are respectively symmetric and anisymmetric under the permutation of 1 and 2,

$$J(1,2) = C^s_{12}J^s(1,2) + C^a_{12}J^a(1,2), \tag{9}$$

where the symmetric part contains the independent emission as in the poisson and the antisymmetric contains information on the gluon interference that break the poisson approximation,

$$|J_2(1,2)|^2_{\text{new}} = \frac{1}{4}[3 + F(1,2)]|J_2(1,2)|^2_{\text{poisson}}. \tag{10}$$

The function $F(1,2)$ is what we call non-abelian correction or gluon interference, this correction depends only on the cross ratio of the four momenta p, q, k_1 ,k_2 respectively the in coming momentum, the tickling momentum, the two radiated momenta. A cross ratio is scale invariant quantity such as

$$\frac{p.k_1 q.k_2}{k_1.k_2 p.q}, \tag{11}$$

which means $F(1,2)$ depends only on the various angle between them, and since the angle between p and q are fixed then it turns out it depends only on three angles

$$F(\theta_1, \theta_2, \phi) = \frac{1 - \cos\theta_1\cos\theta_2}{1 - \cos\theta_1\cos\theta_2 - \sin\theta_1\sin\theta_2\cos\phi}, \tag{12}$$

θ_1 and θ_2 are the emission angles of the emitted gluons while ϕ is the angle between them emission plane, see image below

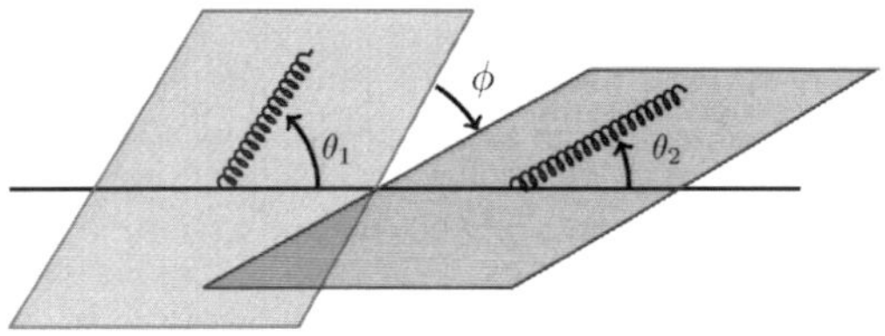

Figure 2: Two gluon emitted in two different plane

To see the effect of the non-abelian correction, let us consider the ratio between the *new* result and the *old* result where it is assumed to be poisson. Then plot them for a fixed $\theta_2 = 0.01$ and ϕ, at first we consider that the gluon are emitted in the same plane $\phi = 0$, then second for different value of ϕ:

- In fig(3a): We can see two different regions, the part in right where $\theta_1 \gg \theta_2$ we have a poissonnian behaviour where there is no interference between the two emission. However when θ_1 approaches the value of θ_2, in the left part, the ratio grows exponentially and diverges at $\theta_1 = \theta_2$.
- In fig(3b): Here we can see that we recover the poisson distribution for $\theta_1 \gg \theta_2$. For $\phi = 0$, in blue, is the same as in fig(3a) where $F(1,2)$ is singular at $\theta_1 = \theta_2$. For $\phi = \frac{\pi}{4}$, in orange, the $F(1,2)$ still grows as θ_1 approaches θ_2 up to some finite value. For $\phi = \frac{\pi}{2}$, in green, $F(1,2)$ becomes constant and equal to 1, i.e. the interference between the two gluons vanishes and the distribution becomes poisson. And last for $\phi = \pi$, in red, $F(1,2)$ decreases as θ_1 approaches the value of θ_2 with a minimal value 0.5 at $\theta_1 = \theta_2$.

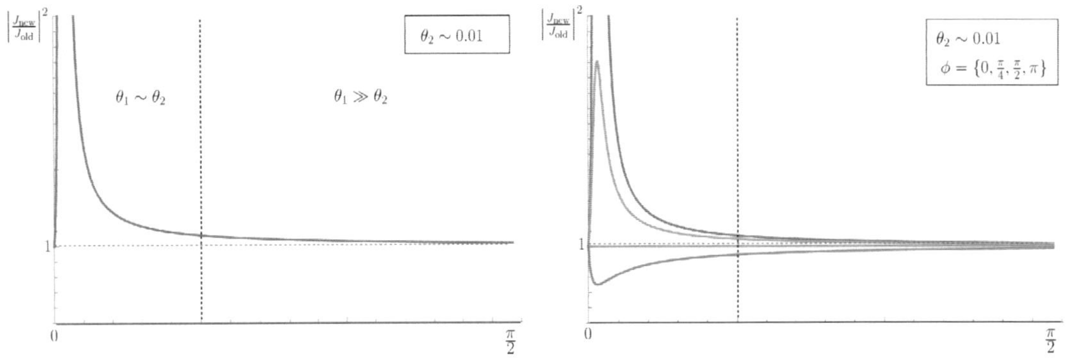

(a) Single Plane emission. (b) ϕ dependence of emission.

Figure 3: plots of the ratio.

5. Conclusion

To conclude let us remind ourselves that we want to improve the momentum distribution for emitting soft-collinear radiation for a quark passing through a QGP. For that we performed a full QCD calculation using the maximal helicity violating techniques in order to avoid diagrammatic complications. And for the two gluon emissions we computed the non-abelian piece $F(1,2)$ which is a correction to the poissonian distribution and it depends only on three variables: the two emission angles and the angle between the emission planes. It has been shown that we can recover the poisson distribution behaviour at some region of the phase space where the angles are strongly ordered or at $\phi = \frac{\pi}{2}$. However outside of these region the interference between gluon become more important and the non-abelian piece (correction) cannot be ignored. Now in order to resum all emission we still need to compute the result for the tree gluons emissions and find more patterns to exponentiate the non-abelian correction.

Acknowledgements

This work would have been impossible without the support of the Science Faculty PhD Fellowship of UCT. We are also grateful to the supports from SA-CERN and NRF-South Africa.

References

[1] Zschiesche, D. et al. *Critical review of quark gluon plasma signals*, Nucl.Phys. A681 (2001) 34-40 nucl-th/0007033 .
[2] Megan Connors, for the ALICE Collaboration, *"Jet Production in p-Pb Collisions"*, arXiv:1409.4655 [nucl-ex].
[3] A. Adare et al., *"Centrality-dependent modification of jet-production rates in deuteron-gold collisions at $\sqrt{s_{NN}}$=200 GeV"*, arXiv:1509.04657 [nucl-ex] (2015).
[4] W.A. Horowitz, *"Heavy Quark Production and Energy Loss"*, Nuclear Physics A, Volumes 904905, Pages 186c193c (2013).
[5] A. H. Mueller, Phys. Lett. 104B (1981) 161; B. I. Ermolayer and V. S. Fadin, JETP Lett. 33 (1981) 269; A. Bassetto, M. Ciafaloni, G. Marchesini and A. H. Mueller, Nucl. Phys. B207 (1982) 189.
[6] Yu. L. Dokshitzer, V. A. Khoze, A. H. Mueller and S. I. Troyan, Rev. Mod. Phys. (1987).
[7] F. Cachazo, P. Svrcek, E. Witten, *"MHV Vertices And Tree Amplitudes In Gauge Theory"*, Journal of High Energy Physics, Volume 2004, JHEP09(2004).
[8] Z. Bern, L. J. Dixon, D. A. Kosower, *"On-shell recurrence relations for one-loop QCD amplitudes"*, Phys. Rev. D 71, 105013, (2005).
[9] A. Brandhuber, B. Spence, G. Travaglini, *"Tree-level formalism"*, Journal of Physics A: Mathematical and Theoretical, Volume 44, Number 45, (2011).

IOP Conf. Series: Journal of Physics: Conf. Series **889** (2017) 012004 doi:10.1088/1742-6596/889/1/012004

Exploring CP-even scalars of a Two Higgs-doublet model in future $e^- p$ colliders

Chuene Mosomane[a,1], Mukesh Kumar[a,2], Alan S. Cornell[a,3] and Bruce Mellado[b,4]

[a] National Institute for Theoretical Physics; School of Physics and Mandelstam Institute for Theoretical Physics, University of the Witwatersrand, Johannesburg, Wits 2050, South Africa.

[b] School of Physics, University of the Witwatersrand, Johannesburg, Wits 2050, South Africa.

E-mail: [1]cjmosomane@gmail.com, [2]mukesh.kumar@cern.ch, [3]alan.cornell@wits.ac.za, [4]bruce.mellado@wits.ac.za

Abstract. In this proceeding we shall explore the potential of a future $e^- p$ collider to probe the CP-even scalars in a two Higgs doublet model. We consider Type-I in this study. The mass of the lighter scalar particle is considered to be the Higgs-boson, $m_h = 125$ GeV, and a heavy scalar $m_H = 270$ GeV. The centre of mass energy for the $e^- p$ collision is considered as in the Large Hadron Electron Collider and the Future Circular Hadron Electron Collider configurations, by fixing the proton beam energy to be $E_p = 7$ and 50 TeV, respectively, and an electron beam energy of $E_e = 60$ GeV. Production cross sections of these scalars are also shown at higher electron beam energies. Future prospects of these studies are also discussed.

1. Introduction

After the discovery and confirmation of the Standard Model (SM) Higgs boson (h) at the Large Hadron Collider (LHC) [1, 2], it is pragmatic to expect hints of any beyond the SM (BSM) particle spectrum by using the present energies in the high luminosity runs. The hints of BSM physics in current and coming data sets from the LHC can be seen in distortions of differential distributions of kinematics of the Higgs boson or any other excesses in data by considering different final states, as noted in Refs. [3, 4, 5]. In these references the authors considered the two-Higgs doublet model (THDM) as a next step in searches of BSM heavy scalars, with a possible spectrum and parameter space at the LHC. In this proceeding we will focus on CP-even scalars h and H of the THDM, and study the possibilities of these scalars at future $e^- p$ colliders, such as the Large Hadron electron Collider (LHeC) and the future circular hadron electron collider (FCC-he).

IOP Conf. Series: Journal of Physics: Conf. Series **889** (2017) 012004 doi:10.1088/1742-6596/889/1/012004

In Section 2 we briefly discuss the THDM, after which we provide a preliminary study on cross section measurements of h and H at the LHeC and FCC-he energies. Future detailed prospects for these studies are pointed out in Section 4.

2. Two-Higgs doublet model

It is well known that in the SM $SU(2)_L \otimes U(1)_Y$ electroweak theory that the Higgs sector consists of only one complex Higgs doublet, and hence only one physical neutral Higgs scalar boson h (where its mass is a free parameter and not fixed by theory). Recall that Run-I and Run-II data at the LHC confirms its mass, $m_h \approx 125$ GeV [1, 2]. If we consider two complex $SU(2)_L$ doublet scalar fields Φ_1 and Φ_2, the Lagrangian for the scalar sector of a THDM can be written as

$$\mathcal{L}_{\Phi_{1,2}} = (D_\mu \Phi_1)^\dagger (D^\mu \Phi_1) + (D_\mu \Phi_2)^\dagger (D^\mu \Phi_2) - V(\Phi_1, \Phi_2), \tag{1}$$

where the covariant derivative in standard notation is given as $D_\mu = \partial_\mu + ig\tau^i/2 \cdot W_\mu^i + ig'/2Y B_\mu$, and the most general renormalisable scalar potential $V(\Phi_1, \Phi_2)$ may be written as:

$$\begin{aligned} V(\Phi_1, \Phi_2) = \; & m_1^2 \Phi_1^\dagger \Phi_1 + m_2^2 \Phi_2^\dagger \Phi_2 - m_3^2 \left(\Phi_1^\dagger \Phi_2 + \text{h.c.}\right) \\ & + \frac{1}{2}\lambda_1 \left(\Phi_1^\dagger \Phi_1\right)^2 + \frac{1}{2}\lambda_2 \left(\Phi_2^\dagger \Phi_2\right)^2 + \lambda_3 \left(\Phi_1^\dagger \Phi_1\right)\left(\Phi_2^\dagger \Phi_2\right) \\ & + \lambda_4 \left|\Phi_1^\dagger \Phi_2\right|^2 + \frac{1}{2}\lambda_5 \left[\left(\Phi_1^\dagger \Phi_2\right)^2 + \text{h.c.}\right] \\ & + \left[\left[\lambda_6 \left(\Phi_1^\dagger \Phi_1\right) + \lambda_7 \left(\Phi_2^\dagger \Phi_2\right)\right] \Phi_1^\dagger \Phi_2 + \text{h.c.}\right]. \end{aligned} \tag{2}$$

After spontaneous breaking of the electroweak symmetry, five physical Higgs particles are left in the spectrum; one charged Higgs pair, $H^\pm$, one CP-odd scalar, A, and two CP-even states, H (heaviest) and h (lightest) given as:

$$H^\pm = \sin\beta\, \phi_1^\pm + \cos\beta\, \phi_2^\pm, \tag{3}$$

$$A = \sin\beta\, \text{Im}\, \phi_1^0 + \cos\beta\, \text{Im}\, \phi_2^0, \tag{4}$$

$$H = \cos\alpha \left(\text{Re}(\phi_1^0) - v_1\right) + \sin\alpha \left(\text{Re}(\phi_2^0) - v_2\right), \tag{5}$$

$$h = -\sin\alpha \left(\text{Re}(\phi_1^0) - v_1\right) + \cos\alpha \left(\text{Re}(\phi_2^0) - v_2\right). \tag{6}$$

Here ϕ_i^+ and ϕ_i^0 denote the $T_3 = 1/2$ and $T_3 = -1/2$ components of the ith doublet for $i = 1, 2$. The angle α diagonalises the CP-even Higgs squared-mass matrix and β diagonalises both the CP-odd and charged Higgs sectors with $\tan\beta = v_2/v_1$, where $\langle \phi_i^0 \rangle = v_i$ for $i = 1, 2$ and $v_1^2 + v_2^2 \approx (246\,\text{GeV})^2$. This is a brief summary of how THDMs are constructed. Furthermore, based on different choices of symmetries, couplings to quarks and leptons etc., different models can be built. Models which lead to natural flavour conservation can be named as Type-I, Type-II, Lepton-specific or Flipped 2HDMs, as detailed in Ref.[6].

Table 1. Models which lead to natural flavour conservation. The superscript i is a generation index.

Models	Type I	Type II	(Leptonic-specific)	(Flipped)
u_R^i	Φ_2	Φ_2	Φ_2	Φ_2
d_R^i	Φ_2	Φ_1	Φ_2	Φ_1
e_R^i	Φ_2	Φ_1	Φ_1	Φ_2

2.1. Yukawa interaction model types

One of the problems we have in the THDM is that we can't diagonalise the mass matrices simultaneously, and this generally cause us to have flavour changing neutral current (FCNCs) at tree level. This means that the Yukawa couplings will not be flavour diagonal. The transformation from a general to a flavour-conserving THDM can be achieved by imposing a discrete symmetry. This follows from the observation that FCNCs vanish at tree level. This results in the THDM having four different model types and can be summarised as in Table 1. We can write the Yukawa Lagrangian that conserves the FCNCs as follows [6]:

$$\begin{aligned}\mathcal{L}_Y = &\bar{Q}_{L,i}(Y^{ij}_{u,1}\tilde{\Phi}_1 + Y^{ij}_{u,2}\Phi_2)u_{R,j} + \bar{Q}_{L,i}(Y^{ij}_{d,1}\Phi_1 + Y^{ij}_{u,2}\tilde{\Phi}_2)d_{R,j} \\ &+\bar{L}_{L,i}(Y^{ij}_{l,1}\Phi_1 + Y^{ij}_{l,2}\Phi_2)l_{R,j} + \text{h.c.},\end{aligned} \tag{7}$$

with $\tilde{\Phi}_i = -i[\Phi_i^\dagger \tau_2]^T, i = 1, 2$ and $Y^{ij}_{u,1}$ being the 3×3 Yukawa matrices. After this brief introduction of THDM, we next present the production cross sections of h and H at the expected energies of future LHeC and FCC-he colliders.

3. Production of h and H at the LHeC and FCC-he

There are various proposals to build new, powerful, high energy e^+e^-, e^-p and pp colliders in the future, where in this proceedings we take an opportunity to probe CP-even scalar Higgs bosons of THDM in e^-p colliders. For our studies we consider the LHeC and FCC-he configurations, where depending on electron beam energies $E_e = 60 - 120$ GeV and the proton beam energy is $E_p = 7$ and 50 TeV respectively. The available centre of mass energies for the LHeC (FCC-he) varies from $\sqrt{s} = 2\sqrt{E_e E_p} \approx 1.3 - 1.8(3.5 - 5.0)$ TeV. For more details we refer to Refs. [7, 8] and more recent updates can be found on the official website [9]. In short these types of colliders

(i) provide a clean environment with suppressed backgrounds from the strong interaction processes, and free from pile-up, multiple interactions etc.,

(ii) have asymmetric initial states, where backward and forward scattering can be disentangled and

(iii) are known for precision measurements of the dynamical properties of the proton.

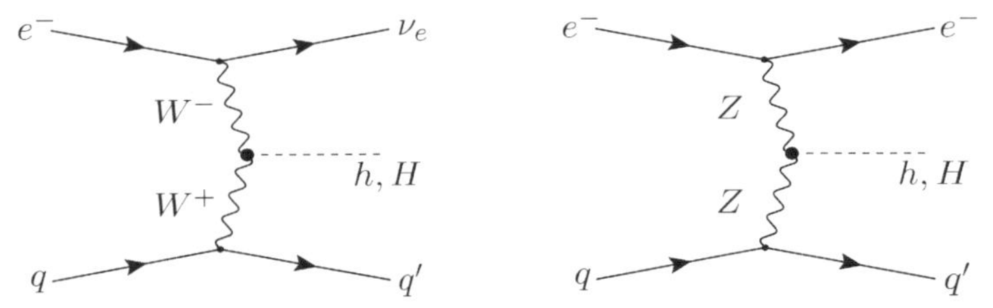

Figure 1. Production modes of h and H through charged-current (left) and neutral current (right) in e^-p colliders in vector-boson fusion modes. Here $q, q' = u, d, c, s, b, \bar{u}, \bar{d}, \bar{c}, \bar{s}, \bar{b}$.

Studies on Higgs-physics in these configurations can be found in Refs. [10, 11, 12, 13]. It is also important to mention that the THDM has been explored in future e^+e^- machines which can be found in the International Linear Collider Higgs white paper [14] and references there-in. In $e^- p$ configurations, h and H can be produced in two ways as shown in Figure 1 (here we only show the vector boson fusion diagrams):

(a) Charged-current (CC): $pe^- \to j\nu_e h/H$ and

(b) Neutral-current (NC): $pe^- \to je^- h/H$,

where $j = u, d, c, s, b, \bar{u}, \bar{d}, \bar{c}, \bar{s}, \bar{b}$. For our studies we use the model files for THDM available in the `SARAH-4.9.3` [15] package, and simulate the events using the Monte Carlo event generator package `MadGraph5-2.4.3.` [16], where we edit the default parameters generated through `SPheno-3.3.8` [17, 18] for our purpose. Before that we also verified the branchings of h and H to all possible decay modes using the `HDECAY-5.10` [19] package. For these calculations we used the `NN23LO1` PDF set.

As a preliminary study we consider CP conserving THDM of Type-I, where the default parameters generated through `SPheno` are only real, and mixings are diagonal. In Figure 2 we have shown the total production cross sections of h and H with respect to electron beam energies by fixing proton beam energies as LHeC and FCC-he recommendations (i.e. 7 and 50 TeV respectively in CC and NC processes where -80% electron polarisation is considered). The parameters are:‡

(a) $m_h = 125$ GeV, $m_H = 270$ GeV, $m_A = 450$ GeV, $m_{H^\pm} = 400$ GeV (as recommended in Refs. [3, 4, 5]),

(b) $\tan\beta = 1.0$, $\alpha = -0.53$ and

(c) $\lambda_1 = 0.1$, $\lambda_2 = 0.27$, $\lambda_3 = 1.1$, $\lambda_4 = -0.5$, $\lambda_5 = 0.5$.

‡ As mentioned in the introduction, the choice of $m_H = 270$ GeV is essential here. In Refs. [3, 4, 5] the authors investigated a number of final states with the 7 and 8 TeV data at ATLAS and CMS. This included the Higgs boson transverse momentum, the limits on the production of di-Higgs bosons, the invariant mass of $VV(V = Z, W^\pm)$ and the results of the search for associated top-Higgs (tth) production. Features in the data could be explained by a hypothetical scalar-boson of mass around 270 GeV.

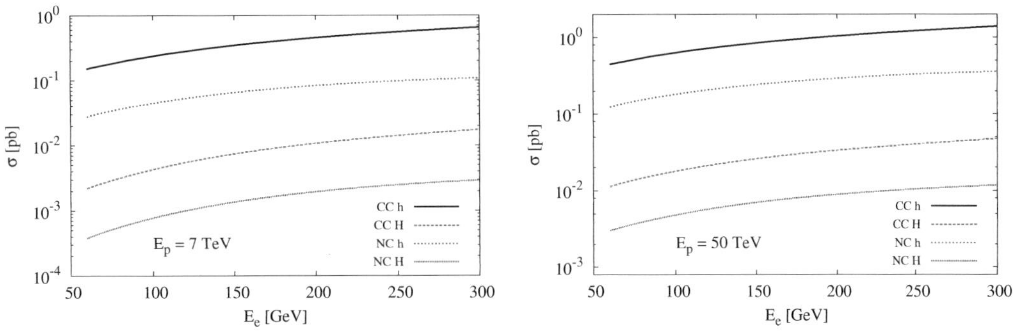

Figure 2. Production cross sections of h and H in charged and neutral current THDM Type-I with respect to electron beam energies E_e for fixed proton beam energy of $E_p = 7$ TeV (left) and $E_p = 50$ TeV (right). The default model parameters are as explained in the text with $\tan\beta = 1.0$ and e^- polarisation is taken as -80%.

Here it is important to mention that the hVV ($V = W^{\pm}, Z$) couplings in the THDM is dependent on the choice of the mixing angles α and β. For example, couplings for hW^+W^-, hZZ depend on $\sin(\beta-\alpha)$, while in the case of HW^+W^-, HZZ, it is proportional to $\cos(\beta-\alpha)$ etc. Note that in the SM there is no such mixing angle dependence on such couplings. In our choice of parameters $\cos(\beta-\alpha) = \pm 0.25$ and $\sin(\beta-\alpha) = 0.97$, which can further be modified and scanned according to the constraints from the ATLAS and CMS analyses at the LHC. Here we chose these values so as to not evade the cross section limits as given in 13 TeV data from ATLAS [20], where a search for new heavy scalar bosons into the four-lepton final state is performed by exploring the mass range [200-1000] GeV. In Table 2 we estimate the branching ratio (BR) of h, H, A and $H^{\pm}$ using the `HDECAY-5.10` package with these parameter choices. The dominant BRs are shown in bold and suggest the possibility for appropriate final states to look for signal to background estimations.

Interestingly one of the important expectations from the LHeC/FCC-he is that we should be able to measure couplings of H with a very good signal to background ratio when compared with pp collisions. In our ongoing work we estimate the appropriate signal to background ratio of H production in CC and NC processes, so that we can have an idea for the level of the BR that can be accessible in these types of machines (which are highly dependent on the THDM parameters). For instance, from Figure 1 for $E_e = 60$ GeV the CC (NC), H production is about 2.2 fb (0.38 fb), which will give us about 2000 (380) H produced for 1 ab^{-1} integrated luminosity at the LHeC. Here it is important to mention that in Ref. [10] the SM Higgs-boson BR to the $b\bar{b}$ mode has been studied, and tagging the forward jet assures this decay mode has a significantly reduced background which improves the purity of the signal. Therefore, estimations of the BRs (shown in Table 2) suggest $H \to hh \to b\bar{b}b\bar{b}$ as the dominant mode (which would be a signal rate of $\sim 55\%$ of the total production cross section of H, where the rate of one of the dominant background $b\bar{b}b\bar{b}j$ in the CC (NC) at the LHeC would be

Table 2. Branching ratio of h, H, A and $H^{\pm}$ by considering the parameter choices as: m_h = 125 GeV, m_H = 270 GeV, m_A = 450 GeV and $m_{H^\pm}$ = 400 GeV, $\tan\beta = 1$, $\alpha = -0.53$, $\lambda_1 = 0.1$, $\lambda_2 = 0.27$, $\lambda_3 = 1.1$, $\lambda_4 = -0.5$ and $\lambda_5 = 0.5$ for THDM Type-I. (Dominant BRs are shown in bold)

Modes	h	Modes	H	Modes	A	Modes	$H^{\pm}$
$\mathbf{b\bar{b}}$	$\mathbf{6.5 \times 10^{-1}}$	$b\bar{b}$	6.8×10^{-4}	$b\bar{b}$	2.7×10^{-4}	bc	5.9×10^{-7}
$\tau^+\tau^-$	$\mathbf{7.0 \times 10^{-2}}$	$\tau^+\tau^-$	8.5×10^{-5}	$\tau^+\tau^-$	3.8×10^{-5}	$\tau\nu$	4.6×10^{-5}
$\mu^+\mu^-$	2.5×10^{-4}	$\mu^+\mu^-$	3.0×10^{-7}	$\mu^+\mu^-$	1.3×10^{-7}	$\mu\nu$	1.6×10^{-7}
$s\bar{s}$	2.5×10^{-4}	$s\bar{s}$	2.6×10^{-7}	$s\bar{s}$	9.6×10^{-8}	su	6.2×10^{-9}
$\mathbf{c\bar{c}}$	$\mathbf{3.2 \times 10^{-2}}$	$c\bar{c}$	3.3×10^{-5}	$c\bar{c}$	1.4×10^{-5}	cs	1.5×10^{-5}
$t\bar{t}$	0.0×10^{-0}	$t\bar{t}$	8.5×10^{-7}	$\mathbf{t\bar{t}}$	$\mathbf{7.6 \times 10^{-1}}$	**tb**	$\mathbf{8.7 \times 10^{-1}}$
gg	$\mathbf{8.5 \times 10^{-2}}$	gg	5.5×10^{-4}	gg	3.1×10^{-3}	cd	8.2×10^{-7}
$\gamma\gamma$	1.4×10^{-3}	$\gamma\gamma$	6.7×10^{-6}	$\gamma\gamma$	9.4×10^{-6}	bu	4.1×10^{-9}
$Z\gamma$	1.0×10^{-3}	$Z\gamma$	1.1×10^{-5}	$Z\gamma$	2.4×10^{-6}	ts	1.4×10^{-3}
$\mathbf{W^+W^-}$	$\mathbf{1.4 \times 10^{-1}}$	$\mathbf{W^+W^-}$	$\mathbf{7.1 \times 10^{-2}}$	**Zh**	$\mathbf{5.1 \times 10^{-2}}$	td	6.5×10^{-5}
ZZ	$\mathbf{1.8 \times 10^{-2}}$	**ZZ**	$\mathbf{3.1 \times 10^{-2}}$	**ZH**	$\mathbf{1.84 \times 10^{-1}}$	$\mathbf{hW^{\pm}}$	$\mathbf{4.6 \times 10^{-2}}$
		hh	$\mathbf{9.0 \times 10^{-1}}$	W^+H^-	3.6×10^{-5}	$\mathbf{HW^{\pm}}$	$\mathbf{8.3 \times 10^{-2}}$

$\sim$ 2.0 fb (22 fb) while for the FCC-he the cross section is around 3 (7.5) times higher for E_e = 60 GeV), while $H \to hh \to W^+W^-$ are subdominant signal modes. Hence the sensitivity of the Hhh, HW^+W^- and HZZ couplings in this mode can be measured at the LHeC/FCC-he, as discussed in Ref. [12].

4. Summary and future work

In this proceedings we have provided a preliminary study of the cross section measurements of h and H using energies from the LHeC and FCC-he configurations. For E_e = 60 GeV and E_p = 7(50) TeV the chosen parameters give substantial signal rates for H production in CC and NC modes (with $m_H = 270$), $\sigma_H^{\rm CC} \sim 2.2(11.3)$ fb and $\sigma_H^{\rm NC} \sim 0.38(3)$ fb respectively. While for h production these rates are $\sigma_h^{\rm CC} \sim 150(450)$ fb and $\sigma_h^{\rm NC} \sim 28(124)$ fb. Here we considered e^- polarisation of -80% which enhances the un-polarised signal cross sections (and also backgrounds) by 1.8 times. We also gave estimations of the BRs for various decay modes of h, H, A and $H^{\pm}$, based on our parameter choice, and discussed its implications. For these parameter choice we discussed dominant final state such as, $H \to hh \to b\bar{b}b\bar{b}$ and thus at the LHeC/FCC-he one can estimate the sensitivity of Hhh couplings including the HW^+W^- and HZZ.

There are various analyses which remain for our future studies, some of these are:

(1) A full analysis of the kinematics of the final states using decay modes of h and

H with appropriate backgrounds and an estimation of the accuracy of the BRs in different final states, say, $b\bar{b}, \tau^+\tau^-$ etc.

(2) A study of the dependence of the parameter space on different observables. It is important to mention here that the choice of parameter space must be consistent with all current Higgs constraints by using `HiggsSignal` and `HiggsBounds` [21]. Though the single parameter choice we took for our studies are also consistent with these constraints.

(3) An exploration of the other scalars of the THDMs, such as the CP-odd A and the charged $H^{\pm}$ should be added for future studies. Note that few studies for light $H^{\pm}$ production and decay in e^-p environment can be found in Refs. [22, 23].

(4) A consistency check with the constrained parameter space of the THDMs through various studies available at the LHC and e^+e^- colliders.

(5) Studies related to the FCNCs, which are also very important.

References

[1] G. Aad *et al.* [ATLAS Collaboration], Phys. Lett. B **716**, 1 (2012)

[2] S. Chatrchyan *et al.* [CMS Collaboration], Phys. Lett. B **716**, 30 (2012)

[3] S. von Buddenbrock *et al.*, arXiv:1506.00612 [hep-ph].

[4] S. von Buddenbrock *et al.*, Eur. Phys. J. C **76**, no. 10, 580 (2016)

[5] M. Kumar *et al.*, J. Phys. Conf. Ser. **802**, no. 1, 012007 (2017)

[6] Branco G. C, Ferreira P. M, Lavoura L, Rebelo M. N, Sher M and Silva J. P, Phys. Rept. **516**, 1 (2012)

[7] J. L. Abelleira Fernandez *et al.* [LHeC Study Group], J. Phys. G **39**, 075001 (2012)

[8] O. Bruening and M. Klein, Mod. Phys. Lett. A **28**, no. 16, 1330011 (2013)

[9] `http://lhec.web.cern.ch`; `http://lhec.web.cern.ch/publications`; `http://lhec.web.cern.ch/talks-seminars`.

[10] T. Han and B. Mellado, Phys. Rev. D **82**, 016009 (2010)

[11] S. S. Biswal, R. M. Godbole, B. Mellado and S. Raychaudhuri, Phys. Rev. Lett. **109**, 261801 (2012)

[12] M. Kumar, X. Ruan, R. Islam, A. S. Cornell, M. Klein, U. Klein and B. Mellado, Phys. Lett. B **764**, 247 (2017)

[13] B. Coleppa, M. Kumar, S. Kumar and B. Mellado, Phys. Lett. B **770**, 335 (2017)

[14] D. M. Asner *et al.*, arXiv:1310.0763 [hep-ph].

[15] F. Staub, arXiv:0806.0538 [hep-ph].

[16] J. Alwall *et al.*, JHEP **1407**, 079 (2014)

[17] W. Porod, Comput. Phys. Commun. **153**, 275 (2003)

[18] W. Porod and F. Staub, Comput. Phys. Commun. **183**, 2458 (2012)

[19] A. Djouadi, J. Kalinowski and M. Spira, Comput. Phys. Commun. **108**, 56 (1998)

[20] The ATLAS collaboration [ATLAS Collaboration], ATLAS-CONF-2016-079.

[21] `http://higgsbounds.hepforge.org` [Date of last access: Tuesday 23 May 2017]

[22] S. Moretti and K. Odagiri, Phys. Rev. D **57**, 5773 (1998)

[23] J. Hernandez-Sanchez, O. Flores-Sanchez, C. G. Honorato, S. Moretti and S. Rosado, arXiv:1612.06316 [hep-ph].

IOP Conf. Series: Journal of Physics: Conf. Series **889** (2017) 012005 doi:10.1088/1742-6596/889/1/012005

Tests with beam setup of the TileCal phase-II upgrade electronics

Dingane Reward Hlaluku

School of Physics, University of the Witwatersrand, Johannesburg 2050, South Africa

E-mail: d.hlaluku@cern.ch

Abstract. The LHC has planned a series of upgrades culminating in the High Luminosity LHC which will have an average luminosity 5-7 times larger than the nominal Run-2 value. The ATLAS Tile calorimeter plans to introduce a new readout architecture by completely replacing the back-end and front-end electronics for the High Luminosity LHC. The photomultiplier signals will be fully digitized and transferred for every bunch crossing to the off-detector Tile PreProcessor. The Tile PreProcessor will further provide preprocessed digital data to the first level of trigger with improved spatial granularity and energy resolution in contrast to the current analog trigger signals. A single super-drawer module commissioned with the phase-II upgrade electronics is to be inserted into the real detector to evaluate and qualify the new readout and trigger concepts in the overall ATLAS data acquisition system. This new super-drawer, so-called hybrid Demonstrator, must provide analog trigger signals for backward compatibility with the current system. This Demonstrator drawer has been inserted into a Tile calorimeter module prototype to evaluate the performance in the lab. In parallel, one more module has been instrumented with two other front-end electronics options based on custom ASICs (QIE and FATALIC) which are under evaluation. These two modules together with three other modules composed of the current system electronics were exposed to different particles and energies in three test-beam campaigns during 2015 and 2016.

1. Introduction

The Large Hadron Collider (LHC) is a proton-proton collider with 14 TeV center of mass energy and design luminosity of $10^{34} cm^{-2} s^{-1}$ [1]. A series of upgrades culminating in the High Luminosity LHC (HL-LHC) have been planned which will result in an increased average luminosity 5-7 times larger than the nominal Run-2 value. The ATLAS detector (A Toroidal LHC Apparatus) [2] is one of the two general purpose particle detectors at the LHC, and it consists of multiple sub-detectors (Figure 1a) which are designed to detect interesting particle physics events in 40 million bunch-crossings per second [3]. The hadronic Tile Calorimeter (TileCal) is the central region of the ATLAS detector and is used to measure energies and directions of hadrons, jets, τ and leptons. Particles pass through the plastic scintillating tiles (Figure 1b) which emit light that is transferred to the photomultiplier tubes (PMTs). This provides the means to measure energy and direction of hadrons. These light signals are directed to the on-detector front-end (FE) electronics where they are converted to electrical signals and digitized, and then processed as first level (L0/L1) trigger events [4]. Data from the FE electronics is further transferred to the off-detector back-end (BE) for further processing and readout to permanent storage. The TileCal FE and BE will be completely redesigned to adopt a new readout strategy to adhere to the HL-LHC parameters [5]. A series of test-beam campaigns during 2015 and 2016 were performed to test and assess the performance of the new prototype electronics

IOP Conf. Series: Journal of Physics: Conf. Series **889** (2017) 012005 doi:10.1088/1742-6596/889/1/012005

for the phase-II upgrades. This article presents the firmware and software developed explicitly for the tests with beam setup and shows preliminary performance analysis for different types of particles and energies as well as for calibration data.

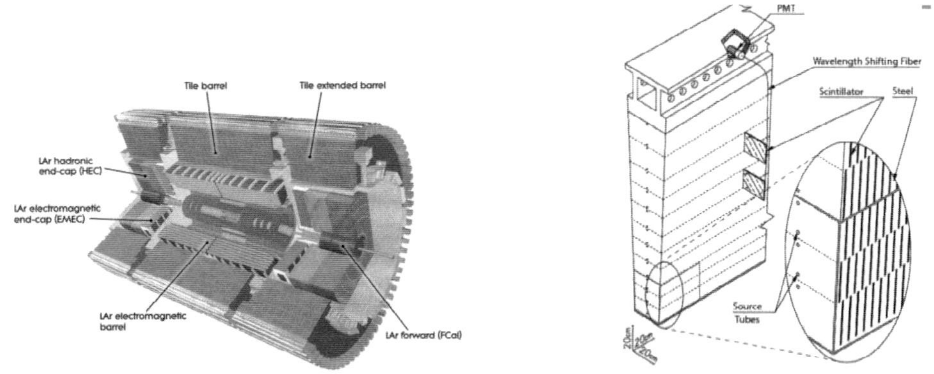

(a) ATLAS Calorimeter system (b) TileCal module

Figure 1: ATLAS Detector and a TileCal module.

2. TileCal Phase-II Upgrade

The LHC has envisaged a series of upgrades in the machine in order to reach the design center of mass energy ($\sqrt{s}$ = 14 TeV) and to increase the peak luminosity up to 5 times the nominal value($\mathcal{L}_{peak}$ = 5 x $10^{34} cm^{-2} s^{-1}$) in 2023. The TileCal detector components (absorber, scintillating tiles, fibers, and PMTs) are largely in good shape and do not have to be replaced. The readout electronics are to be completely redesigned to provide full-granularity digital data to the Level-0/1 triggers at 40 MHz. The present on-detector electronics are designed to output digital data at the maximum rate of about 100 kHz with the digital data stored on detector in a 6.4 μs long pipeline (Figure 2), and are basically not compatible with the HL-LHC TDAQ (Trigger and Data Acquisition) architecture [6]. Most of the components have aged (time & radiation) and some of them have even been discontinued. The main difference between the present and the upgrade readout architectures is that all the digital data is transmitted off-detector at 40 MHz (Figure 3). The data is pipelined and processed in the off-detector Tile PPr (PreProcessor).

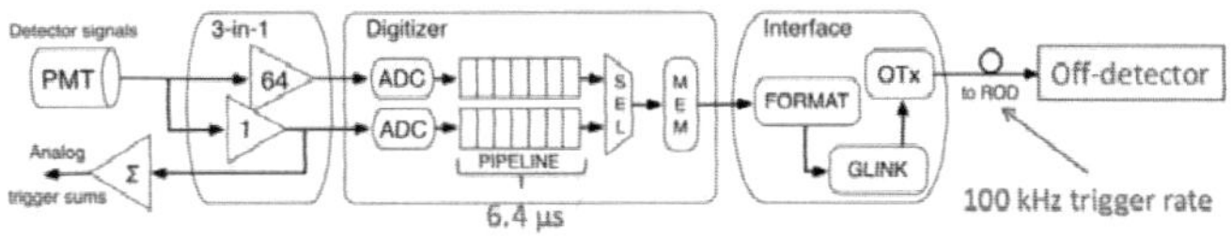

Figure 2: Current readout architecture.

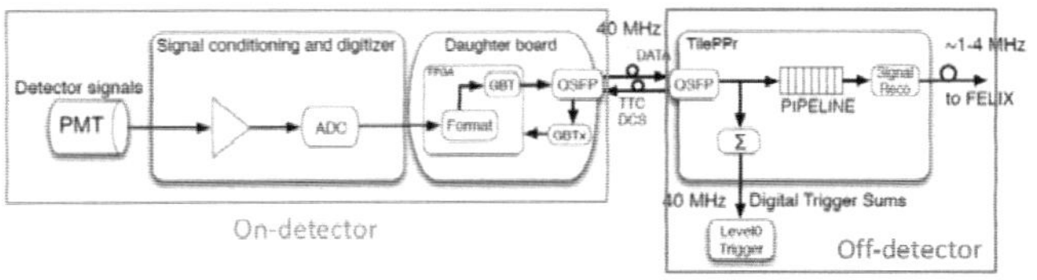

Figure 3: Upgrade readout architecture.

The Demonstrator project aims at testing the long term performance of the upgrade system without compromising the present data taking. This is to be achieved by adding to the upgrade design the analog trigger signal output. Digital transformation will translate TTC (Timing, Trigger and Control) signals into upgrade commands and translate upgrade outputs to a format acceptable by the present RODs (Read Out Drivers). The super-drawer Demonstrator module is composed of four independent mini-drawers (Figure 4), each of them are equipped with:

- 12 FE boards: 1 out of the 3 different options.
- 1 Main Board (MB): for the corresponding FE option.
- 1 Daughter-board (DB): single design.
- 1 HV (High Voltage) regulation board: 1 out of 2 options viz. Remote HV and HVOpto.
- 1 adder base board plus 3 adder cards for the summation of the analog trigger signals to send to the L0/1 event accept.
- New LVPS (Low Voltage Power Supply) architecture that provides redundancy and Point Of Load regulators - LVPS v.8.01 [7].

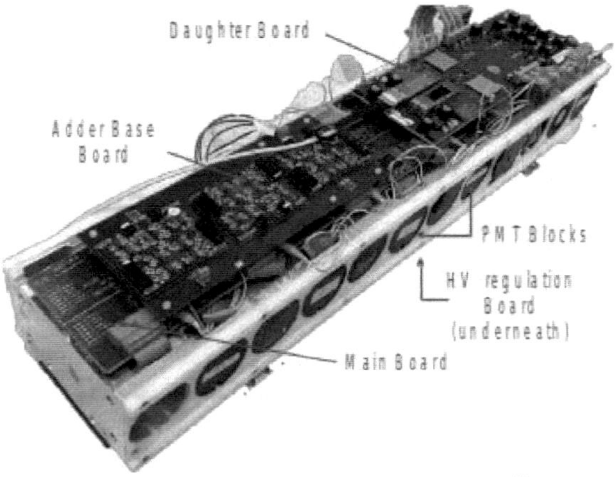

Figure 4: The Demonstrator mini-drawer.

3. Test-beam campaigns

Three test-beam campaigns took place in 2015 and 2016 with the aim of testing the phase-II upgrade electronics prototypes and to assess their performance. Several TileCal modules were equipped with the phase-II upgrade electronics and exposed to different beam particles (electrons, muons and hadrons) during these test-beam campaigns. The primary objective was to assess the status of the Demonstrator based on the modified 3-in-1 FE baseline option, attention and help was also given to the two alternative FE options viz. the QIE (Charge Integrator and Encoder) [8] and the FATALIC (Front-end ATLas TIle Circuit.) [9]. The results from these campaigns provide the necessary performance analysis to aid the 2017 FE option down-selection. The module configuration for the test-beam campaigns is depicted in Figure 5 below.

There is new hardware components available as of the 2016 test-beams, this include the DB version 4 which uses the new GBTx (GigaBit Transceiver) chip [10] instead of the CDCE chip for clock recovery and remote FPGA (Field Programmable Gate Arrays) programming [11, 12]. The MB version 2 which provides a frame and bit clock for ADC de-serialization, correct channel orientation and new 3-in-1 cards. The PPr prototype with no auxiliary boards and operates at 40.08 MHz (4,8 Gbps - 9,6 Gbps) with optimized firmware versions. The PCle (Peripheral Component Interconnect Express) module for

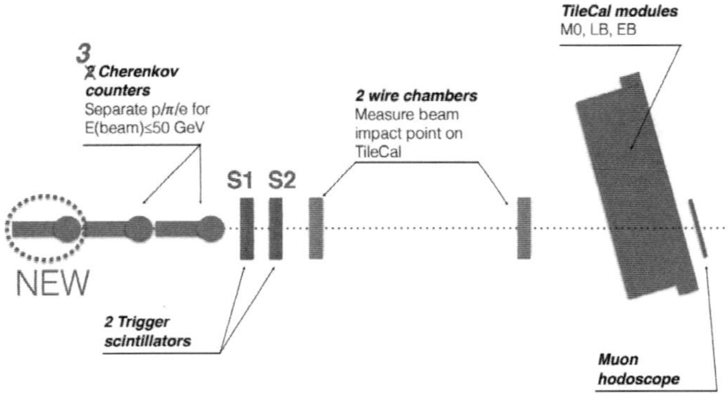

Figure 5: Test-beam configuration with TileCal modules.

readout to the basic pseudo-FELIX(Front-End LInk eXchange) [13] through the PCle-GBT link data format. A new mobile Cs-137 calibration unit to perform in-situ Cesium calibrations. The 3 FE options under evaluation were all operated in parallel for the entire test-beam period. New software for the TDAQ and the DCS (Detector Control Systems) to improve the communication through IPbus with PPr was developed and provides; improved PPr memory stability for multiple application access, monitoring the link status and FE option configuration at the beginning of every data taking session. The DCS software has been updated to work with the official IPbus OPC server [14] that provides better HV and temperature monitoring. Calibration runs (CIS, CIS mono, Ped and LED) have been successfully integrated as well.

The 3-in-1 cards (Figure 6a) are built with discrete commercial off-the-shelf components. The shaped pulses from 12 FE cards are digitized on the MB. The upgrade brings an improved digitized signal resolution: 12 bit ADCs instead of 10 bits, modern, better performing components. It also brings even lower electronic noise than the present system. The new 3-in-1 system is compatible with the analog trigger and can be installed into TileCal before the HL-LHC upgrade. As mentioned before, the new 3-in-1 system was calibrated and evaluated with test beams hadrons, muons (Figure 6b), and electrons (Figure 6c) in 2015 and 2016. The results illustrate a strong performance for the upgrade system.

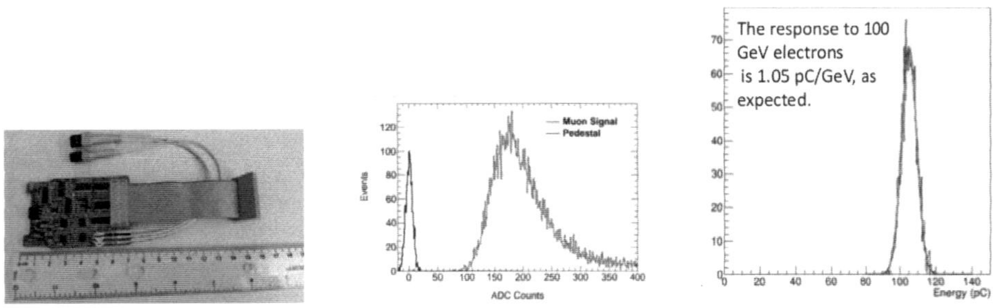

(a) 3-in-1 FE board. (b) Muon signal response. (c) Electron signal response.

Figure 6: Sample Test-beam analysis results for the 3-in-1

The FATALIC (Figure 7a) is a new technology for TileCal, it is implemented as an AISC (Application Specific Integrated Circuit) slightly different from the 3-in-1. The FATALIC approach also uses shaping like the 3-in-1 but the pulse shape is different. The digitization is done inside the chip. Simulated performance for FATALIC_5 IC is within specifications. Work is ongoing on signal reconstruction and simulations. The FATALIC system was calibrated and evaluated with test beams hadrons, muons (Figure 7b), and electrons

(Figure 7c) in 2016. The results also speak in favor of the strong performance of the FATALIC system.

(a) FATALIC FE board.

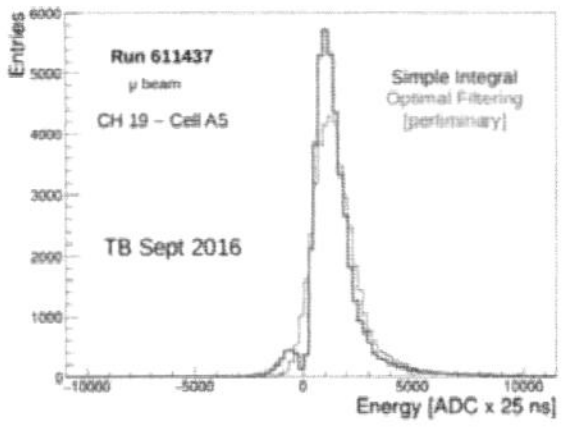

(b) Muon signal response.

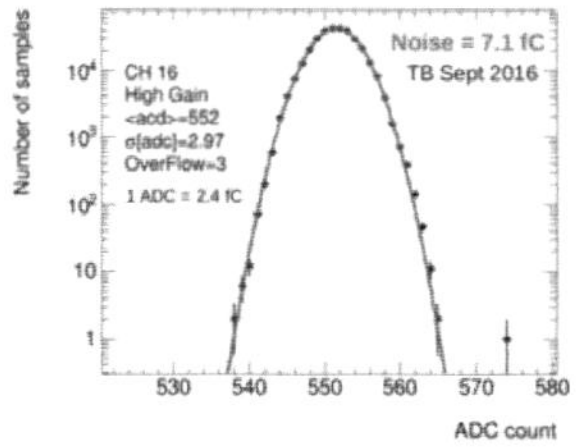

(c) Response from electron signals.

Figure 7: Sample Test-beam analysis results for the FATALIC

The QIE circuit (Figure 8a) is an ASIC which does not shape the PMT pulse to digitization at 40 million samples per second. Instead, it directly integrates the PMT anode current in 25 ns intervals and gives two sample signals viz. (i) charge (current integrated over a 25 ns period) and (ii) time (when current crossed a threshold at 1 ns resolution). The total charge of a PMT pulse (energy) is obtained as a sum of two or three QIE samples. The PMT current is integrated in a bank of capacitors that are time-multiplexed for lossless operation at 40 MHz. The current splitter is used to achieve the required dynamic range (>17 bits). The QIE12 ASIC is designed to be radiation tolerant and tailored for TileCal. All the signal processing (integration and digitization) is done in the FE boards. QIE main boards have mostly point-to-point signal connections and power distribution [15]. The QIE system was calibrated and evaluated with test beams hadrons, muons (Figure 8b), and electrons (Figure 8c) in 2016. The results also speak in favor of the strong performance of the QIE system.

(a) QIE FE board.

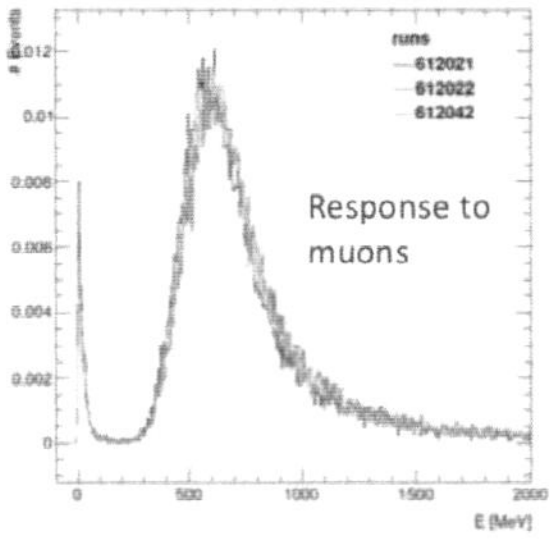

(b) Muon signal response.

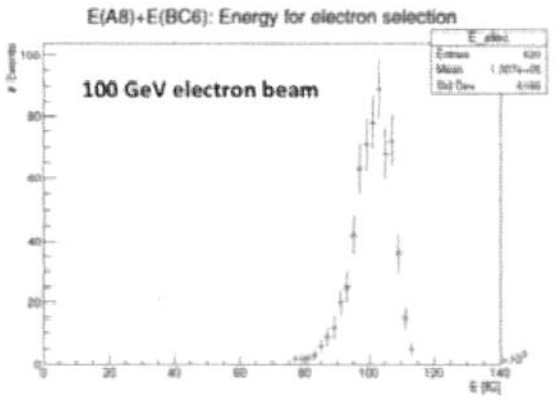

(c) Response from electron signals.

Figure 8: Sample Test-beam analysis results for the QIE12

4. Conclusions

Upgrade activities are progressing well with the goal of test-beams to do performance analysis for the 2017 option down-selection. The reliability and stability of the system has been visibly improved with respect to the 2015 test-beam. The FE-BE links have been optimized for better stability, improved latency and FE clock recovery. The Demonstrator analog trigger signals output has been integrated successfully into the TDAQ infrastructure for the commissioning of the hybrid Demonstrator. The FE and HV option final selection process has been defined based on extensive comparison of the performances and other criteria. The TileCal community plans to complete the R&D and take a decision by the end of summer.

IOP Conf. Series: Journal of Physics: Conf. Series **889** (2017) 012005 doi:10.1088/1742-6596/889/1/012005

References

[1] The ATLAS Collaboration. *ATLAS detector and physics performance: Technical Design Report, 1.* Technical Design Report ATLAS. CERN, Geneva, 1999.

[2] The ATLAS Collaboration. The ATLAS Experiment at the CERN Large Hadron Collider. *J. Instrum.*, 3:S08003. 437 p, 2008. Also published by CERN Geneva in 2010.

[3] The ATLAS Collaboration. Letter of Intent for the Phase-II Upgrade of the ATLAS Experiment. Technical Report CERN-LHCC-2012-022. LHCC-I-023, CERN, Geneva, Dec 2012. Draft version for comments.

[4] Matthew Spoor and Oscar Kureba and Charles Sandrock. Development of a high data-throughput ADC board for the PROMETEO portable test-bench for the upgraded front-end electronics of the ATLAS TileCal. *Journal of Physics: Conference Series*, 645(1):012027, 2015.

[5] F Carrio and V Castillo and A Ferrer and L Fiorini and Y Hernndez and E Higon and B Mellado and L March and P Moreno and R Reed and C Solans and A Valero and J A Valls. The sROD module for the ATLAS Tile Calorimeter Phase-II Upgrade Demonstrator. *Journal of Instrumentation*, 9(02):C02019, 2014.

[6] O. Solovyanov. Performance of the ATLAS Tile Hadronic Calorimeter at LHC in Run 1 and planned upgrades. *Journal of Instrumentation*, 9(10):C10006, 2014.

[7] ATLAS Tile Collaboration. Initial Design for the Phase-II Upgrade of the ATLAS Tile Calorimeter System. Technical Report ATL-COM-TILECAL-2016-054, CERN, Geneva, Dec 2016.

[8] Gary Drake and Alexander Paramonov. Controls and read-out of QIE12 front-end cards. Technical Report ATL-COM-TILECAL-2016-043, CERN, Geneva, Oct 2016.

[9] Laurent Royer. FATALIC: A Dedicated Front-End ASIC for the ATLAS TileCal Upgrade. Oct 2015.

[10] P Moreira, S Baron, S Bonacini, O Cobanoglu, F Faccio, S Feger, R Francisco, P Gui, J Li, A Marchioro, C Paillard, D Porret, and K Wyllie. The GBT-SerDes ASIC prototype. *Journal of Instrumentation*, 5(11):C11022, 2010.

[11] Kai Chen, Hucheng Chen, Weihao Wu, Hao Xu, and Lin Yao. Optimization on fixed low latency implementation of GBT protocol in FPGA. Technical Report arXiv:1608.08916, CERN, Aug 2016.

[12] S Muschter, C Soos, C Bohm, J-P Cachemiche, and S Baron. Optimizing latency in Xilinx FPGA implementations of the GBT. *J. Instrum.*, 5:C12017, 2010.

[13] J T Anderson et al. FELIX: a PCIe based high-throughput approach for interfacing front-end and trigger electronics in the ATLAS Upgrade framework. Technical Report ATL-DAQ-PROC-2016-022. 12, CERN, Geneva, Nov 2016.

[14] Soare Cristian-Valeriu. OPC UA IPbus server. Sep 2015.

[15] Gary Drake, Alexander Paramonov, James Proudfoot, Robert Stanek, and Sergei Chekanov. QIE12: A New High-Performance ASIC for the ATLAS TileCal Upgrade. *ATLAS*, Oct 2015.

IOP Conf. Series: Journal of Physics: Conf. Series **889** (2017) 012006 doi:10.1088/1742-6596/889/1/012006

What Can Gamma-rays from Space tell us About the Madala Hypothesis?

Geoff Beck and Sergio Colafrancesco

School of Physics, University of the Witwatersrand, Private Bag 3, WITS-2050, Johannesburg, South Africa

E-mail: geoffrey.beck@wits.ac.za, sergio.colafrancesco@wits.ac.za

Abstract. The recent Madala hypothesis, a conjecture that seeks to explain anomalies within Large Hadron Collider (LHC) data (particularly in the transverse momentum of the Higgs boson), is interesting for more than just a statistical hint at unknown and unpredicted physics. This is because the model itself contains additional new particles that may serve as Dark Matter (DM) candidates. These particles interact with the Standard Model via a scalar mediator boson S. More interesting still, the conjectured mass range for the DM candidate (65 - 100 GeV) lies within the region of models viable to try explain the recent Galactic Centre (GC) gamma-ray excess seen by Fermi Large Area Telescope (Fermi-LAT) and the High Energy Stereoscopic System (HESS). Therefore, assuming S decays promptly, it should be possible to check what constraints are imposed upon the effective DM annihilation cross-section in the Madala scenario by hunting signatures of S decay that follows DM annihilation within dense astrophysical structures. In order to make use of existing data, we use the Reticulum II dwarf galaxy and the galactic centre gamma-ray excess data sets from Fermi-LAT, and compare these to the consequences of various decay paths for S in the aforementioned environments. We find that, based on this existing data, we can limit τ lepton, quark, direct gamma-ray, and weak boson channels to levels below the canonical relic cross-section. This allows us to set new limits on the branching ratios of S decay, which can rule out a Higgs-like decay branching for S, in the case where the Madala DM candidate is assumed to comprise all DM.

1. Introduction

The unknown nature of dark matter (DM) sits as a major eye-sore on the completeness of the concordance cosmology. Multiple methods are being pursued in order to catch a glimpse of this elusive substance. In particular, collider searches like at the LHC ATLAS [1] and CMS [2] experiments can probe numerous models of theoretical interest [3, 4, 5]. However, there is one promising candidate to be found within the Madala hypothesis [6, 7]. This hypothesis was put forward to explain anomalies in existing LHC data but also conveniently provides a DM candidate (see Section 3 for details). This candidate is especially interesting because kinematic considerations place it in the mass range 65 to 100 GeV [7]. This region is significant as it lies within collection of models that might be compatible with explaining the gamma-ray excess seen in the galactic centre [8, 9] by Fermi-LAT [10] and HESS [11], as well as anti-particle excesses, according to the consideration of modelling uncertainties performed by [12].

For these reasons we will examine the consequences of the annihilation of DM introduced by the Madala hypothesis within both the galactic centre and the promising dwarf galaxy target Reticulum II. Using Fermi-LAT gamma-ray data from these targets [13, 14] we can formulate limits on possible annihilation cross-sections for multiple channels connecting the Madala DM to the Standard Model by comparing the existing data to gamma-ray fluxes expected from DM annihilation within these astrophysical structures. In addition to this we will examine the scalar S, that mediates with the DM particle in the Madala scenario, comparing what limits we can place on the decay branching ratios of S to the assumption S is Higgs-like [7]. This will be performed under the assumption that Madala DM constitutes all DM in the universe (and thus has an annihilation cross-section set to the canonical relic value [15]).

This paper is structured as follows: in Section 2 we provide more detail on the galactic centre gamma-ray excess as well as observations of Reticulum II. In Section 3 we will detail the relevant aspects of the Madala hypothesis and its bearing on the problem of DM. In Section 4 we will detail the formalism used to calculate gamma-ray emmisions from GC and Reticulum II. Finally, in Section 5 we present and discuss our results.

2. The Galactic Centre Gamma-ray Excess and Reticulum II

The galactic centre has been a prime target for DM hunts in gamma-rays ever since the discovery of unexpected gamma-ray excesses around 1 - 10 GeV [8, 9]. This is evinced by the numerous works dedicated to using DM to explain the excess [16, 17, 18, 19, 20, 21, 22]. In our work here we will make use of the spectrum for the excess within the region of interest (ROI) between 1° and 20° from the galactic centre found by [14]. This choice of ROI is a standard when analysing galactic centre gamma-ray data (see [14] and references therein) in order to avoid the emissions of the powerful Sagittarius A* complex within the galactic core region.

Reticulum II is a faint dwarf galaxy recently found by the Dark Energy Survey (DES) project [23, 24]. It is notable because it is calculated to posses a very large J-factor [25], a parameter that tracks the density of dark matter within halos. The density is abetted by the fact that Reticulum II is very close to Earth (30 kpc away [24]), thus reducing inverse-square flux attenuation. This particular dwarf was the source of some speculation, as there appeared a small excess in its gamma-ray spectrum observed by Fermi-LAT [26] (attributed by the aforementioned authors at 2σ confidence level to a WIMP with relic cross-section and around 60 GeV mass). However, subsequent Fermi collaboration analysis revealed that the excess could not associated with DM when considered against other dwarf galaxy targets [13]. This was reinforced by an analysis in [27], which concluded that the Reticulum II DM model would produce unacceptable excesses in the radio and gamma-ray spectra of other targets. Despite this, Reticulum II makes an ideal test-bed for producing constraints on DM models, as a large J-factor means large DM-induced fluxes from annihilation, which can then be compared to the Fermi-LAT upper limits. We make use of the upper limits established on gamma-ray fluxes from Reticulum II as used in [13].

3. Madala Hypothesis and Dark Matter

The Madala hypothesis is one that sets out to explain anomalies in LHC data concerning the transverse momentum of the Higgs boson within LHC collisions (among others - see [6, 7]). It does so through the introduction of a set of particles: a large scalar "Madala" boson H (~ 270 GeV mass) which is Higgs-like, and a scalar S with a mass range 130 - 200 GeV which couples a dark particle χ to the standard model and to H. This scalar S acts as mediator between the standard model and dark particles proposed in the Madala hypothesis (as seen in Fig. 1). Therefore, in order to explore what kinds of indirect astrophysical signatures might be expected from DM that results from this conjecture, one must concentrate upon the possible couplings, and thus decay paths, between S and the Standard Model. Such limits may be of particular interest in determining which couplings to the standard model are permissible for S given current astrophysical data.

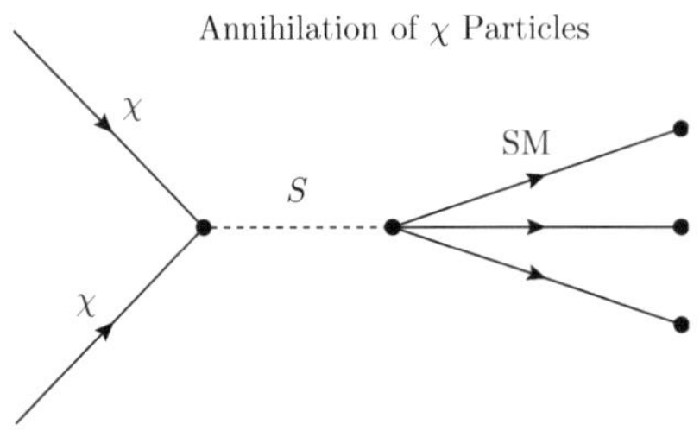

Figure 1. A diagram of χs annihilating into S which promptly decays to some SM particles.

4. Dark Matter halos and gamma-ray flux

In a given DM halo, the differential gamma-ray flux resulting from annihilation can be specified by

$$\phi(E_\gamma, \Delta\Omega, l) = \frac{1}{4\pi}\frac{\langle\sigma V\rangle}{2m_\chi^2}\frac{dN_\gamma}{dE_\gamma}\int_{\Delta\Omega}\int_l \rho^2(\mathbf{r})dl'd\Omega' \ , \tag{1}$$

where E_γ is the gamma-ray energy, m_χ is the mass of the WIMP, ρ is the DM halo spatial density profile, $\langle\sigma V\rangle$ is the velocity averaged thermal annihilation cross-section, and $\frac{dN_\gamma}{dE_\gamma}$ is the γ-ray yield from S decay following WIMP annihilations (sourced from PYTHIA [28] routines in DarkSUSY [29] as well as [30, 31]). In this work we will assume S decays promptly, thus we will place a limit on the effective annihilation cross-section for the process $\chi\chi \to S \to SM$.

The expression Eq. (1) can be simplified by splitting it into two factors. The first is the astrophysical "J"-factor, which encompasses the above two integrals,

$$J(\Delta\Omega, l) = \int_{\Delta\Omega}\int_l \rho^2(\mathbf{r})dl'd\Omega' \ , \tag{2}$$

with the integral being extended over the line of sight l, and $\Delta\Omega$ is the observed solid angle. The second factor is determined only by particle physics:

$$\psi(E_\gamma) = \frac{1}{4\pi}\frac{\langle\sigma V\rangle}{2m_\chi^2}\frac{dN_\gamma}{dE_\gamma} \ . \tag{3}$$

Thus the flux will be found from

$$\phi(E_\gamma) = \psi(E_\gamma) \times J(\Delta\Omega, l) \ . \tag{4}$$

For the Reticulum II dwarf we take the J-factor to be 2.0×10^{19} GeV^2 cm^{-5} [25]. However, for the galactic centre we follow the methodology of [14] and use a contracted NFW profile [32]

$$\rho_N(r, r_s, \eta) = \frac{\rho_s}{\left(\frac{r}{r_s}\right)^\eta \left(1 + \frac{r}{r_s}\right)^{3-\eta}} \ , \tag{5}$$

where we take the scale radius $r_s = 20$ kpc following [14], and ρ_s is defined by ensuring that $\rho = 0.4$ GeV cm^{-3} at a radius of 8.5 kpc. We then calculate the average J-factor for a profile with $\eta = 1.2$ for the ROI between 1° and 20° (as explained in Section 2) from the galactic centre using formula Eq. (2). We choose this particular profile as contracted NFW halos are often found in best-fit scenarios for DM explanations of the GC excess [14, 12, 19], thus making these results more relevant to the existing literature.

5. Results and Discussion

In Figure 2 we display the limits derived from gamma-ray fluxes in the galactic centre and Reticulum II on the effective annihilation cross-section into a variety of standard model particles (assuming a branching fraction of $b_f = 1$ for each channel individually). The most significant region of the plot is the purple Madala mass band, where the DM masses correspond with the mass range expected for the boson S. Here we see that, for the channels $\tau^+\tau^-$, $q\bar{q}$, and direct gamma-ray production, we can use both targets to explore the region of the parameter space in which the Madala DM constitutes all the DM in the universe (relic band). In the case of Reticulum II we can also do this for the Higgs channel. For the weak boson channels we can probe below the relic band of cross-sections only in Reticulum II. We note that neither of these targets can be used to rule out the entire region of the Madala band that overlaps the galactic centre excess region from [12]. However, we stress that Reticulum II has only extant upper-limits on its gamma-ray flux, thus we might expect these already strong constraints to improve with further observations.

The significance of these limits can be understood as follows: if the cross-section can be constrained below the relic level then we rule out the DM model as a candidate for all DM, as its present abundance would be too great to match cosmological constraints. However, since we assume $b_f = 1$ in each case, we can instead derive a limit on the decay branching for S should the Madala DM particle constitute all DM (assuming $\langle \sigma V \rangle = 3 \times 10^{-26}$ cm^3 s^{-1} for all channels). The results of this analysis are shown in Table 1. Any entry with a dash signifies that no constraint can be derived (we leave out the light leptons as this is true for all masses). The final column shows the branching ratios for the Standard Model Higgs boson with mass equal to S in each case (as S is assumed Higgs-like [7]).

What is evident from these results is that both decay of S into quarks and into gamma-rays as well as W and Z bosons (not shown on table) can be constrained below $\sim 30\%$. For W bosons this means that the branching of S rules out the Higgs-like case (which ranges from ~ 0.5 to ~ 0.75 in the suggested S mass range [33]). While for $q\bar{q}$ our constraint rules out the Higgs-like range for S masses below $\lesssim 145$ GeV. Our constraints for direct photon and Z decays cannot rule out a Higgs-like S. Limits on decay of S into Higgs bosons show large variability with WIMP mass. This arises from the movement of the gamma-ray resonant peak within the data domain for each source. Due to the hardness of the resulting gamma-ray spectrum [27], the τ lepton channel is subject to weaker constraints, with $\sim 70\%$ limit being possible for the largest masses considered, thus this channel does not affect the Higgs-like S case. The branching ratios of S into light leptons cannot be constrained at all.

This leaves considerable room for a Madala WIMP to constitute all DM, but raises serious doubts about a Higgs-like S if the Madala hypothesis is to account for $> 50\%$ of DM. Finally, future radio frequency searches with the Square Kilometre Array [34] can probe significantly lower cross-sections [27] (even in the presence of large radio background fluxes) and will thus be integral in further constraint, or dismissal, of the Madala hypothesis and its attendant particles.

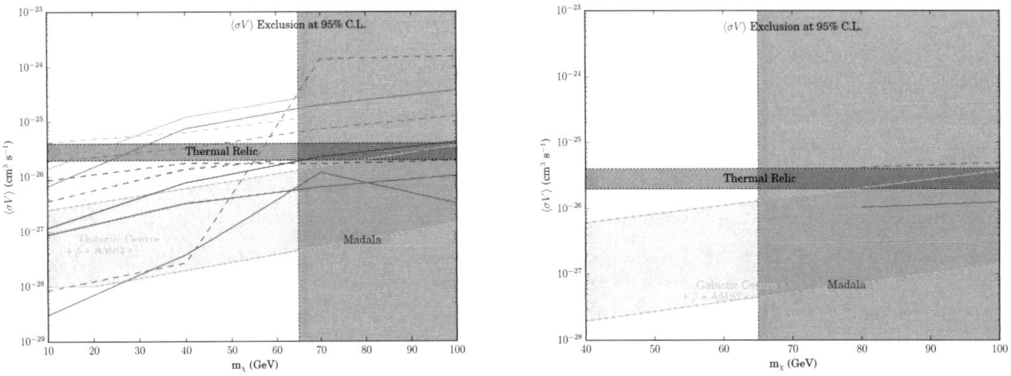

Figure 2. Constraints on effective annihilation cross-section as a function of WIMP mass for various channels shown in comparison to galactic centre excess model region (pink). Reticulum II constraints are shown as solid lines while galactic centre results are shown as dashed lines. Top: $q\bar{q}$ in yellow, Higgs in green, e^+e^- in red, $\tau^-\tau^-$ in blue, $\mu^+\mu^-$ in orange, and direct gamma-rays in magenta. Bottom: W^+W^- in green and Z bosons in yellow.

Channel	m_χ	$\langle\sigma V\rangle_f$ (GC)	b_f (GC)	$\langle\sigma V\rangle_f$ (Ret. II)	b_f (Ret. II)	b_f [33]
$q\bar{q}$	40	1.49e-26	0.5	3.37e-27	0.11	0.87
$q\bar{q}$	70	2.38e-26	0.79	6.46e-27	0.22	0.33
$q\bar{q}$	100	-	-	1.03e-26	0.34	2.5e-3
h	40	3e-28	0.01	3.79e-28	0.013	-
h	70	-	-	1.24e-26	0.41	-
h	100	-	-	3.22e-27	0.11	-
$\tau^+\tau^-$	40	1.8e-26	0.6	7.79e-27	0.26	8.3e-2
$\tau^+\tau^-$	70	1.72e-26	0.57	2.27e-26	0.76	3.5e-2
$\tau^+\tau^-$	100	1.97e-26	0.66	-	-	2.9e-4
γ	40	1.36e-26	0.45	3.37e-27	0.11	9.2e-4
γ	70	2.27e-26	0.76	6.46e-27	0.22	1.9e-3
γ	100	-	-	1.03e-26	0.34	5.5e-5

Table 1. Table showing constraints on Madala DM for various masses and channels derived from the galactic centre (GC) and Reticulum II (Ret. II). Both a constraint for $\langle\sigma V\rangle$ assuming $b_f = 1$ is shown as well as a constraint on the decay branching when it is assumed $\langle\sigma V\rangle$ takes the value of the relic cross-section. This b_f can be compared against the Standard Model Higgs results for the appropriate mass choice [33].

IOP Conf. Series: Journal of Physics: Conf. Series **889** (2017) 012006 doi:10.1088/1742-6596/889/1/012006

Acknowledgments
S.C. acknowledges support by the South African Research Chairs Initiative of the Department of Science and Technology and National Research Foundation and by the Square Kilometre Array (SKA). G.B acknowledges support from a post-doctoral grant through the same initiative and institutions.

References
[1] G. Aad et al., JINST, 3, S08003 (2008).
[2] M. Della Negra, A. Petrilli, A. Herve, & L. Foa, *CMS Physics Technical Design Report Volume I: Software and Detector Performance* (2008), `http://doc.cern.ch//archive/electronic/cern/preprints/lhcc/public/lhcc-2006-001.pdf`
[3] H. M. Lee, M. Park, & W. Park, Phys. Rev., D, 86, 103502 (2012).
[4] J Brooke, M. R. Buckley, P. Dunne, B. Penning, J. Tamanas, & M. Zgubič, Phys. Rev. D, 93, 113013 (2016).
[5] F. Bishara & J. J. Zupan, JHEP, 1601, 010 (2016).
[6] S. von Buddenbrock, N. Chakrabarty, A. S. Cornell, D. Kar, M. Kumar, T. Mandal, B. Mellado, B. Mukhopadhyaya, & R. G. Reed, preprint, arXiv:1506.00612 [hep-ph] (2015).
[7] S. von Buddenbrock, N. Chakrabarty, A. S. Cornell, D. Kar, M. Kumar, T. Mandal, B. Mellado, B. Mukhopadhyaya, R. G. Reed, & X. Ruan, Eur. Phys. J. C, 76, 580 (2016), arXiv:1606.01674 [hep-ph].
[8] M. Ackermann et al. for the Fermi-LAT collaboration, Phys. Rev. D, 82, 092004 (2010). arXiv: 1008.3999.
[9] F. Aharonian et al. for the Fermi-LAT collaboration, Astron. Astrophys. 425, L13 (2004).
[10] W. B. Atwood et al. for the Fermi-LAT collaboration, Astrophys. J., 697, 1071 (2009). arXiv:0902.1089 [astro-ph].
[11] `https://www.mpi-hd.mpg.de/hfm/HESS/`
[12] F. Calore, I. Cholis, C. McCabe, & C. Weniger, Phys. Rev. D, 91, 063003 (2015), arXiv:1411.4647 [astro-ph]
[13] A. Drlica-Wagner et al. for the Fermi-LAT collaboration & T. Abbott et al. for the DES collaboration, ApJ, 809, L4 (2015), arXiv: 1503.02632 [astro-ph].
[14] T. Daylan, D. P. Finkbeiner, D. Hooper, T. Linden, S. K. N. Portillo, N. L. Rodd, & T. L. Slatyer, Physics of the Dark Universe, 12, 1 (2016). arXiv: 1402.6703.
[15] G. Jungman, M. Kamionkowski, & K. Griest, J. Phys. Rep., 267, 195 (1996).
[16] L. Goodenough & D. Hooper, preprint, arXiv:0910.2998 (2009).
[17] D. Hooper & L. Goodenough, Phys. Lett. B, 697, 412 (2011), arXiv:1010.2752.
[18] A. Boyarsky, D. Malyshev, & O. Ruchayskiy, Phys. Lett. B, 705, 165 (2011), arXiv:1012.5839.
[19] D. Hooper & T. Linden, Phys. Rev. D, 84, 123005 (2011), arXiv:1110.0006.
[20] K. N. Abazajian & M. Kaplinghat, Phys. Rev. D, 86, 083511 (2012), arXiv:1207.6047.
[21] C. Gordon & O. Macias, Phys. Rev. D, 88, 083521 (2013), arXiv:1306.5725.
[22] K. N. Abazajian, N. Canac, S. Horiuchi, & M. Kaplinghat, Phys. Rev. D, 90, 023526 (2014), arxXiv:1402.4090.
[23] `http://www.darkenergysurvey.org`
[24] K. Bechtol et al. for the DES collaboration, ApJ, 807(1), 50 (2015), arXiv:1503.02584 [astro-ph.GA].
[25] V. Bonnivard, C. Combet, D. Maurin, A. Geringer-Sameth, S. M. Koushiappas, M. G. Walker, M. Mateo, E. Olszewski, & J. I. Bailey III, ApJ, 808, L36 (2015).
[26] A. Geringer-Sameth, M. G. Walker, S. M. Koushiappas, S. E. Koposov, V. Belokurov, G. Torrealba, & N. W. Evans, Phys. Rev. Lett., 115, 081101 (2015), arXiv:1503.02320 [astro-ph].
[27] G. Beck & S. Colafrancesco, JCAP, 05, 013 (2016).
[28] T. Sjöstrand, Comput. Phys. Commun., 82, 74 (1994).
[29] P. Gondolo, J. Edsjo, P. Ullio, L. Bergstrom, M. Schelke, & E.A. Baltz, JCAP, 0407, 008 (2004).
[30] M. Cirelli, G. Corcella, A. Hektor, G. Htsi, M. Kadastik, P. Panci, M. Raidal, F. Sala, & A. Strumia, JCAP, 1103, 051 (2011). Erratum: JCAP, 1210, E01 (2012). arXiv 1012.4515.
[31] P. Ciafaloni, D. Comelli, A. Riotto, F. Sala, A. Strumia, & A. Urbano, JCAP, 1103, 019 (2011). arXiv 1009.0224.
[32] J. F. Navarro, C. S. Frenk, & S. D. M. White, ApJ, 462, 563 (1996).
[33] S. Heinemeyer (ed) et al. for the LHC Higgs Cross Section Working Group, *Handbook of LHC Higgs Cross Sections: 3. Higgs Properties : Report of the LHC Higgs Cross Section Working Group* (CERN, 2013). arXiv: 1307.1347 [hep-ph].
[34] P. Dewdney, W. Turner, R. Millenaar, R. McCool, J. Lazio, & T. Cornwell, *SKA baseline design document* (2012), `http://www.skatelescope.org/wp-content/uploads/2012/07/SKA-TEL-SKO-DD-001-1_BaselineDesign1.pdf`.

IOP Conf. Series: Journal of Physics: Conf. Series **889** (2017) 012007 doi:10.1088/1742-6596/889/1/012007

Quasi normal modes for a spin-3/2 field in a 4-dimensional Reissner-Nordström black hole background.

X. Ngcobo[1], **A. S. Cornell**[2] **and G. E. Harmsen**[3]

National Institute for Theoretical Physics; School of Physics and Mandelstam Institute for Theoretical Physics, University of the Witwatersrand, Johannesburg, Wits 2050, South Africa

E-mail: [1]xolane.ncgobo@students.wits.ac.za, [2]alan.cornell@wits.ac.za and [3]gerhard.harmsen5@gmail.com

Abstract. We will present a theoretical framework for studying Quasi Normal Modes (QNMs) for spin-3/2 fields in a 4-dimensional Reissner-Nordström black hole background, as a foundation for understanding higher-dimensional black holes. This work is useful in cases where we wish to study the signatures of micro higher-dimensional black holes, and how we might detect them during the experiments performed in the LHC at CERN. We shall make use of the Wentzel-Kramers-Brilloin (WKB) method to compute the QNMs for black holes perturbed by fields. We take the approximation to 6th order and compare these results to those we obtain using the improved asymptotic iterative method. Using the WKB method to 3rd order we are able to generate absorption probabilities associated to our spin-3/2 fields. Finally we compare these results to those obtained for the Schwarzschild black hole.

1. Introduction

Quasi Normal Modes (QNMs) are a result of perturbations of black holes and they can give an insight into the stability of the black hole [1, 2]. The black hole perturbations considered here are due to spin-3/2 fields. In this work we will be comparing Quasi Normal Frequency (QNF) results we obtained for a spin-3/2 field in a 4-dimensional Reissner-Nordström background using the Wentzel-Kramers-Brilloin (WKB) and Asymptotic Iterative Method (AIM) to those obtained for the Schwarzschild black hole. Using the equations of motion we are able to derive a radial equation which has the form

$$\frac{d^2\Psi}{dr^2} + \left(\omega^2 - V(r)\right)\Psi = 0\,. \tag{1}$$

This Schrödinger like form is necessary in order to apply the chosen approximation methods. The QNMs are given by ω and $V(r)$ is the potential felt by the field near the black hole. At the boundaries of our space time we require that our wave functions behave like plane waves. To ensure this we assume our functions are of the following form:

$$\begin{aligned} \Psi &\propto e^{i\omega r_*} \quad \text{for} \quad r_* \to \infty, \\ \Psi &\propto e^{-i\omega r_*} \quad \text{for} \quad r_* \to -\infty, \end{aligned} \tag{2}$$

where r_* is the tortoise coordinate given as $dr_*/dr = 1/f(r)$, with $1/f(r)$ the function for the radial component of the metric. In order to derive our equations of motion we use the massless form of the Rarita-Schwinger equation [3, 4]:

$$\gamma^{\mu\nu\rho}\mathcal{D}_\nu\Psi_\rho = 0\,, \tag{3}$$

where in Eq. (3) $\gamma^{\mu\nu\rho}$ is the anti-symmetric gamma tensor and is given as

$$\gamma^{\mu\nu\rho} = \gamma^{[\mu}\gamma^\nu\gamma^{\rho]} = \gamma^\mu\gamma^\nu\gamma^\rho - \gamma^\mu g^{\nu\rho} + \gamma^\nu g^{\mu\rho} - \gamma^\rho g^{\mu\nu}\,, \tag{4}$$

where our γ^μ's denote the gamma matrices. Our metric is given as [5]

$$ds^2 = -f(r)dt^2 + \frac{dr^2}{f(r)} + r^2 d\Omega^2\,, \tag{5}$$

with $f(r) = 1 - 2M/r + Q/r^2$, with Q the charge of the black hole and M the mass of the black hole, and $d\Omega^2 = \sin^2\theta\, d\theta^2 + d\phi^2$. Where the location of our event horizons are determined by $r = 1 \pm \sqrt{M^2 - Q^2}$, as such for a physical solution we must have $Q \le M$. The case of $Q = M$ represents the extremal case for our black hole, which we will investigate and compare these results to those obtained for the Schwarzschild black hole given in Ref. [6]. We hope that by considering the extremal case we will identify any significant differences in our new results. Solving for the equations of motion as done in Ref. [6] we achieve the following radial equation

$$-\frac{d^2}{dr_*^2}\bar{\phi}_1 + V_1\bar{\phi}_1 = \omega^2\bar{\phi}_1; \;\; -\frac{d^2}{dr_*^2}\bar{\phi}_2 + V_2\bar{\phi}_2 = \omega^2\bar{\phi}_2, \tag{6}$$

where r_* is the tortoise coordinate given as $d/dr_* = fd/dr$, $V_{1,2} = \pm f(r)dW/dr + W^2$, and

$$W = \frac{\sqrt{f}}{ABr}\left[\bar{\lambda}AB + CAB + \bar{\lambda} + C - \bar{\lambda}f\right]\,, \tag{7}$$

with

$$C = \frac{Q}{r}; \qquad A = \sqrt{f} + \bar{\lambda} + \frac{Q}{r} \qquad \text{and} \qquad B = \sqrt{f} - \bar{\lambda} - \frac{Q}{r}.$$

In the above expressions $\bar{\phi}_1$ and $\bar{\phi}_2$ are the wave functions associated to our fields, and $\bar{\lambda}$ are the eigenvalues for our spinors on a sphere given as $\bar{\lambda} = j + 1/2$ with $j = 3/2, 5/2, ...$ [6].

In Fig. 1 we have plotted the effective potential for the extremal Reissner-Nordström black hole and the Schwarzschild black hole, where the results for the uncharged Schwarzschild black hole are given in Ref. [6]. We observe that in both cases the potential function increases with an increase in the value of j. However there are some interesting differences between our two cases, firstly the Reissner-Nordström black hole has its event horizon closer to the singularity, implying it is more compact than the Schwarzschild black hole. Secondly, the Reissner-Nordtröm black hole has a greater potential maximum, for all equivalent values of j, so the black hole should be producing more energetic QNMs.

Once we have obtained the effective potential for our spin-3/2 particles we may begin applying approximation methods to determine our allowed QNMs, this is is outlined in the next section. In section 3 we present these results and use them to determine the absorption probabilities associated to our black holes in section 4. Finally we provide some concluding remarks and outline future works.

IOP Conf. Series: Journal of Physics: Conf. Series **889** (2017) 012007 doi:10.1088/1742-6596/889/1/012007

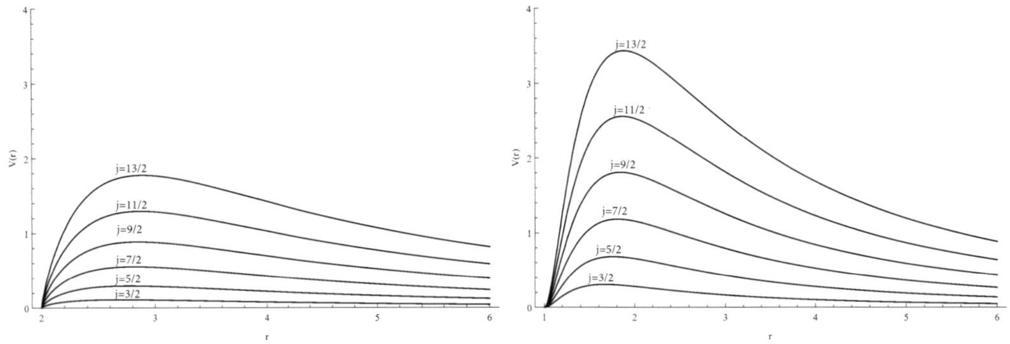

(a) $Q = 0$ (b) $Q = M$

Figure 1: QNMs for both the Schwarzschild and Reissner-Norström black holes in 4 dimensions for values of $j = 3/2$ to $j = 13/2$.

2. Approximation methods

2.1. WKB

The WKB approximation method is a semi-classical approach to solving Schrödinger-like equations. The method was first used to study black hole perturbations due to scalar fields, and later the first order WKB method was used to study QNMs associated to black hole perturbations [1, 7]. The method was extended to 6th order by R. A. Konoplya where he showed that the QNMs can be obtained using the following equation [7]

$$\frac{i\left(\omega^2 - V_0\right)}{\sqrt{-2V_0''}} - \sum_{i=2}^{6} \Lambda_i = n + \frac{1}{2}, \tag{8}$$

where the Λ_i are correction terms, and can be found in Ref. [8]. V_0 and V_0'' are evaluated at the point r_0 which represents the radial location of our maximal value for the potential. So once we have determined the values of V_0 and V_0'' we may solve for ω, which represents our complex QNMs.

2.2. Improved AIM

The improved AIM method can be used to solve second order differential equations, such as our radial equation. The original AIM used an iterative process, in which we must take a derivative in each step to obtain a solution for the second order differential equation. This is computationally time consuming as such the new method uses instead a recursion relation which does away with the derivative operators. As we have already explained this method in a previous proceedings we refer the reader to Ref. [9, 10] for a full explanation of the method.

3. QNMs

In Tab. 1 we observe that the results for the AIM and the WKB to 6th order agree very closely, where as those for the WKB 3rd order do not have a strong agreement with the other two results. This strong agreement in the 6th order WKB and the AIM suggests that our results are correct. We note that the higher modes for each value of j have an increased dampening term. This means that if we were to detect QNMs from a Reissner-Nordström black we are most likely to detect the lower n modes and least likely to detect the higher modes. In Fig. 2 we

IOP Conf. Series: Journal of Physics: Conf. Series **889** (2017) 012007 doi:10.1088/1742-6596/889/1/012007

Table 1: Low-lying ($n \leq l$, with $l = j - 3/2$) spin-3/2 field quasi normal mode frequencies using the WKB and the AIM methods in a 4-dimensional Reissner-Nordström black hole background.

l	n	WKB 3rd order	WKB 6th order	AIM
0	0	0.5410-0.0867i	0.5414-0.0865i	0.5414-0.0864i
1	0	0.8173-0.0874i	0.8174-0.0874i	0.8174-0.0873i
1	1	0.8027-0.2638i	0.8032-0.2636i	0.8032-0.2636i
2	0	1.0810-0.0878i	1.0811-0.0878i	1.0811-0.0877i
2	1	1.0701-0.2643i	1.0704-0.2643i	1.0703-0.2642i
2	2	1.0491-0.4435i	1.0491-0.4435i	1.0490-0.4435i
3	0	1.3395-0.0880i	1.3396-0.0880i	1.3395-0.0879i
3	1	1.3307-0.2646i	1.3309-0.2646i	1.3308-0.2645i
3	2	1.3136-0.4430i	1.3136-0.4430i	1.3135-0.4430i
3	3	1.2889-0.6239i	1.2880-0.6245i	1.2880-0.6245i
4	0	1.5953-0.0881i	1.5953-0.0881i	1.5953-0.0881i
4	1	1.5879-0.2648i	1.5880-0.2648i	1.5880-0.2648i
4	2	1.5734-0.4427i	1.5734-0.4427i	1.5734-0.4427i
4	3	1.5523-0.6225i	1.5517-0.6228i	1.5517-0.6228i
4	4	1.5252-0.8047i	1.5232-0.8060i	1.5232-0.8060i
5	0	1.8494-0.0882i	1.8494-0.0882i	1.8494-0.0882i
5	1	1.8430-0.2649i	1.8431-0.2649i	1.8431-0.2649i
5	2	1.8305-0.4425i	1.8305-0.4425i	1.8305-0.4425i
5	3	1.8121-0.6216i	1.8117-0.6218i	1.8117-0.6218i
5	4	1.7883-0.8026i	1.7869-0.8034i	1.7869-0.8034i
5	5	1.7596-0.9857i	1.7563-0.9879i	1.7563-0.9879i

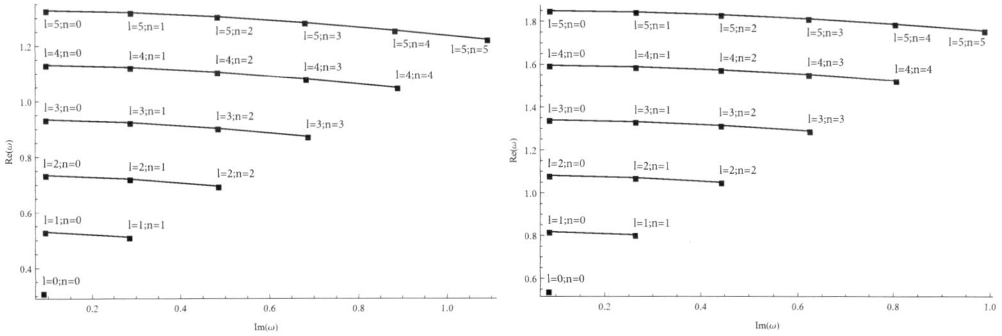

(a) $Q = 0$ (b) $Q = M$

Figure 2: QNMs for both the Schwarzschild and Reissner-Norström black holes in 4 dimensions for values of $l = 0$ to $l = 5$.

have plotted a comparison between the Schwarzschild black hole and the Reissner-Nordström black hole, using the results of Ref. [6]. Comparing our results we see that our QNFs are indeed larger and therefore more energetic for the Reissner-Nordström black hole. We also see that the dampening terms for the Reissner-Nordström black hole have not increased as drastically as the QNFs.

IOP Conf. Series: Journal of Physics: Conf. Series **889** (2017) 012007 doi:10.1088/1742-6596/889/1/012007

4. Absorption probabilities

In order to relate our QNMs to cross sections we need to determine the absorption probabilities associated with the QNMs associated with the Reissner-Norsdtröm black hole. The method and theory for determining the absorption probabilities is given in Ref. [1, 4, 6, 11]. Due to the high energy of some of our emitted fields we need to use the WKB to 3rd order approach to determine our absorption probabilities. The formula is given as

$$|A_j(\omega)|^2 = \frac{1}{1+e^{2S(\omega)}}\,, \tag{9}$$

where

$$\begin{aligned} S(\omega) &= \pi k^{1/2}\left[\tfrac{1}{2}z_0^2 + \left(\tfrac{15}{64}b_3^2 - \tfrac{3}{16}b_4\right)z_0^4\right] + \pi k^{-1/2}\left[\tfrac{3}{16}b_4 - \tfrac{7}{64}b_3^2\right] \\ &\quad + \pi k^{1/2}\left[\tfrac{1155}{2048}b_3^4 - \tfrac{315}{256}b_3^2 b_4 + \tfrac{35}{128}b_4^2 + \tfrac{35}{64}b_3 b_5 - \tfrac{5}{32}b_6\right] z_0^6 \\ &\quad - \pi k^{-1/2}\left[\tfrac{1365}{2048}b_3^4 - \tfrac{525}{256}b_3^2 b_4 + \tfrac{85}{128}b_4^2 + \tfrac{95}{64}b_3 b_5 - \tfrac{25}{32}b_6\right] z_0^2, \end{aligned} \tag{10}$$

and

$$z = r - r_0\,;\quad z_0^2 \equiv -2\frac{P_0}{P_0''}\,;\quad k \equiv \frac{1}{2}P_0''\,; \tag{11}$$

$$b_n \equiv \left(\frac{2}{n!P_0''}\right)\left(\frac{d^nP}{dr_*^n}\right)_0\,;\quad \frac{d}{dr_*} = \left(1 - \frac{2M}{r} + \frac{Q}{r^2}\right)\frac{d}{dr}\,, \tag{12}$$

with $P = \omega - V$ [6]. Subscript 0's denote evaluations at r_0, the radial location of the maximum of our potential. Plugging the determined values for our QNMs into this equation gives us the results as shown in Fig. 3.

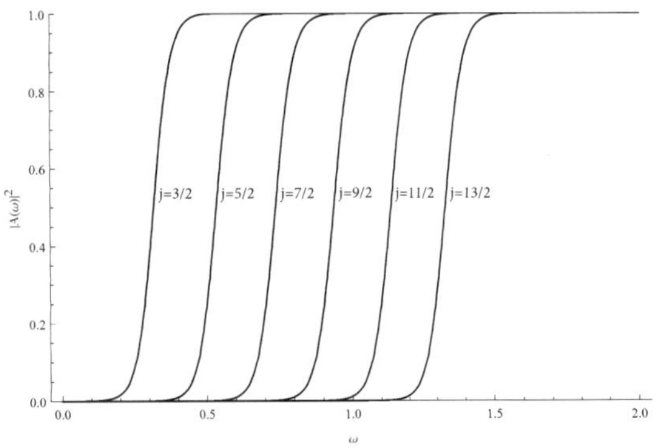

(a) $Q = 0$

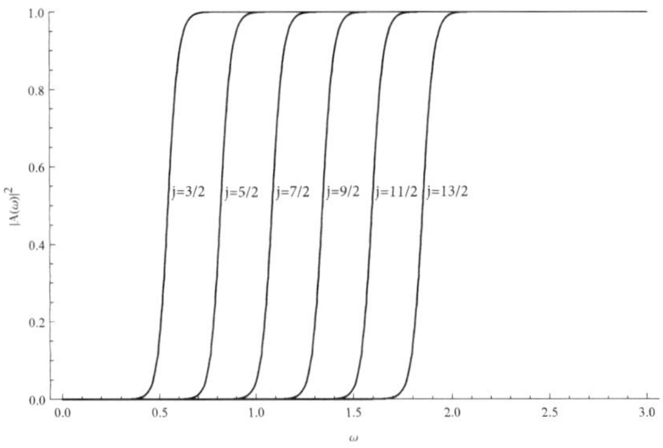

(b) $Q = M$

Figure 3: Absorption probabilities for both the Schwarzschild and Reissner-Norström black holes in 4 dimensions for values of $j = 3/2$ to $j = 13/2$.

In Fig. 3 we have plotted the values for our absorption probabilities associated to the Schwarzschild and the Reissner-Nordström black hole. We notice that increase in the value of j results in an increase in the minimum energy required for total absorption to occur. It is also clearly visible that total absorption occurs at higher energies for the Reissner-Nordström black hole.

IOP Conf. Series: Journal of Physics: Conf. Series **889** (2017) 012007 doi:10.1088/1742-6596/889/1/012007

5. Concluding Remarks

The allowed QNMs for our spin-3/2 particles near our extremal Reisssner-Nordstrom black hole are given in Tab. 1, where we can clearly see that an increase in the value of quantum number l results in an increase in the frequency of our emitted QNMs. Furthermore we see that an increase in n results in an increase in the dampening term of our QNMs. These results are consistent with those obtained in Ref. [6], comparing our result to those in this paper we also note that the Reissner-Nordström black hole emits QNMs with a higher energy when compared to the Schwarzschild black hole.

In Fig. 3 we have plotted the absorption values of our fields for values of $j = 3/2$ to $j = 13/2$. We note that an increase in the value of j results in an increase in the minimum energy required for total absorption to occur, these results are again consistent to the results we have obtained in Ref. [6]. We also note that the minimum energy required for total absorption to occur in the Reissner-Nordström black hole is higher than the Schwarzschild black hole.

In this paper we have considered only the 4-dimensional Reissner-Nordström case, in a future paper we will extend this to higher dimensional Reissner-Nordström black holes. Having done this we will use the same method and tools of analysis to determine the allowed QNMs for the Anti-de Sitter type black hole.

Acknowledgements

We would like to thank Prof. Hing-Tong Cho and Dr. Chun-Hung Chen, at the Tamkang university, for their help and input during this research.

References

[1] Iyer S and Will C M 1987 *Phys. Rev. D* **35**(12) 3621–3631
[2] Kokkotas K D and Schmidt B G 1999 *Living Rev. Rel.* **2** 2
[3] Rarita W and Schwinger J 1941 *Phys. Rev.* **60**(1) 61–61
[4] Chen C H, Cho H T, Cornell A S, Harmsen G E and Ngcobo X *To be published*
[5] Liu J T, Pando Zayas L A and Yang Z 2014 *Journal of High Energy Physics* **2014** ISSN 10298479
[6] Chen C H, Cho H T, Cornell A S and Harmsen G 2016 *Phys. Rev.* **D94** 044052
[7] Konoplya R A and Zhidenko A 2011 *Rev. Mod. Phys.* **83** 793–836
[8] Konoplya R 2003 *Physical Review D* **68** 024018
[9] Ciftci H, Hall R L and Saad N 2003 *Journal of Physics A: Mathematical and General* **36** 11807
[10] Harmsen G E 2015 Quasi-normal modes for spin-3/2 fields *Journal of Physics: Conference Series* vol 645 (IOP Publishing) p 012003
[11] Cho H and Lin Y 2005 *Classical and Quantum Gravity* **22** 775

IOP Conf. Series: Journal of Physics: Conf. Series **889** (2017) 012008 doi:10.1088/1742-6596/889/1/012008

A generalization to all system sizes of DGLV energy loss in the quark-gluon plasma

Isobel Kolbé[1] and W. A. Horowitz[1]

[1]Department of Physics, University of Cape Town, Private Bag X3, Rondebosch 7701, South Africa

E-mail: `isobel.kolbe@gmail.com`

Abstract. In the last few years, experiments with small colliding systems (such as $p/d + A$) have yielded some evidence for the presence of small droplets of quark-gluon plasma (QGP). The collective behavior, strangeness enhancement and quarkonium suppression that have been seen beg the question: What of the energy loss? However, in order to answer such a question, one must apply existing perturbative Quantum Chromodynamical (pQCD) energy loss calculations to systems that are much smaller than the systems in which pQCD energy loss calculations were originally so successful, a process that is not currently allowed for in any pQCD energy loss calculation. We relax the large system assumption in the well-known DGLV formalism (an assumption that is also utilized in the ASW, BDMPS-Z, AM and HT calculations), but find, alarmingly, that the correction terms dominate at large energies, resulting in $\sim$ 100% *negative* correction, calling into question the validity of the large formation time assumption used in *all* pQCD-based energy loss calculations. Our results demand a complete overhaul of energy-loss calculations for all system sizes.

1. Introduction

In recent years, alarming questions have been raised concerning the creation of a quark-gluon plasma (QGP) in small colliding systems such as proton or deuteron collisions with heavy nuclei ($p/d + A$) or even proton-proton (pp) collisions. A number of signals (such as collective behavior [1], strangeness enhancement [2, 3], and quarkonium suppression [4]) that are traditionally attributed to the presence of a QGP in nucleus-nucleus, AA, collisions, have appeared in systems previously thought far too small to create a QGP. During the very successful studies of the QGP in heavy ion collisions, jet quenching proved an important femtoscopic probe of the dynamics of the degrees of freedom of the QGP and tantalizing signs of centrality dependent jets in p/dA collisions at RHIC and LHC [5, 6] have surfaced.

Jet tomography in heavy-ion collisions has been extremely successfully described by a number of perturbative Quantum Chromodynamical (pQCD) energy loss models [7–10] that are able to qualitatively describe the momentum dependence and angular distribution of the suppression of high-momentum, $\sim 5 - 150$ GeV single particle pions [11, 12] and charged hadrons [13–15] from primordial hard light flavors and gluons and electrons [16–18] as well as D [19] and non-prompt J/ψ mesons [20] from open heavy flavor decays at mid rapidity in A+A systems from $\sqrt{s} = 0.2$ TeV to 2.76 TeV.

The presence of some QGP signals in small colliding systems demands a better understanding of energy loss in small systems, since *all* of the standard pQCD energy loss models make the

IOP Conf. Series: Journal of Physics: Conf. Series **889** (2017) 012008 doi:10.1088/1742-6596/889/1/012008

explicit assumption that the QGP system is large [21]. In the standard opacity expansion developed by Djordjevic, Gyulassy, Levai and Vitev (DGLV) [22,23], the large system assumption amounts to an assumption that the separation distance $\Delta z \equiv z_1 - z_0 \gg \lambda_{mfp} \gg 1/\mu$ between the initial production position z_0 of the hard parent parton and the position z_1 where it scatters off a QGP medium quasi-particle is large. The mean free path of the high-p_T particle is $\lambda_{mfp} = 1/\rho\sigma \sim 1-2$ fm while the Debye mass in an infinite, static thermal QGP of temperature $T \sim 350$ MeV is $\mu = gT \sim 0.5$ GeV, as derived from thermal field theory [24]. In pp and p/dA collisions, one expects a system of radius $\lesssim 2$ fm, suggesting that a high-p_T parton that does scatter cannot have a large separation distance between production and scattering.

In order to apply DGLV energy loss to small systems, we derive the $N = 1$ opacity generalization for all separation distances. We find two curious results:

(i) In addition to being suppressed under the large separation distance (along with the usual assumptions of eikonality and collinearity), the majority of the small separation distance terms are also suppressed under the assumption that the gluon formation time is large, which results in only two diagrams with non-zero correction terms.

(ii) The (negative) correction terms dominate even for relatively large system sizes at high parent parton energies.

It has been known for some time that all energy loss formalisms are sensitive to the collinear approximation [25,26], but we will demonstrate that the present sensitivity to the large formation time approximation is both new and different.

2. Setup

Our computation is available in full detail at [27] and follows DGLV [23] which considers a brick of QGP being traversed by a high-p_T eikonal parton that had been produced at an initial point $(t_0, z_0, \mathbf{x}_0)$. In our notation we have described 2D vectors using $\mathbf{p}$, 3D vectors using $\vec{\mathbf{p}} = (p_z, \mathbf{p})$ and four vectors in Minkowski and light cone coordinates respectively using $p = (p^0, \vec{\mathbf{p}}) = [p^0 + p^z, p^0 - p^z, \mathbf{p}]$. In keeping with DGLV, we describe the scattering potention using a Gyulassy-Wang Debye screened potential [28] with Fourier and color structure given by

$$\begin{aligned} V_n &= V(\vec{\mathbf{q}}_n) e^{-i\vec{q}_n \cdot \vec{x}_n} \\ &= 2\pi\delta(q^0) v(\mathbf{q}_n, q_n^z) e^{-i\vec{q}_n \cdot \vec{x}_n} T_{a_n}(R) \otimes T_{a_n}(n). \end{aligned} \tag{1}$$

The $SU(N_c)$ generator $T_a(n)$ in the d_n dimensional representation of the target or $T_a(R)$ in the d_R dimensional representation of the high-p_T parent parton manage the color exchanges. The momenta, defined in Fig. 1, of the radiated gluon, the final high-p_T parton, and the exchanged medium Debye quasi-particle are, in lightcone coordinates,

$$k = \left[xP^+, \frac{m_g^2 + \mathbf{k}^2}{xP^+}, \mathbf{k}\right], p = \left[(1-x)P^+, \frac{M^2 + \mathbf{k}^2}{(1-x)P^+}, -\mathbf{k}\right], q = [q^+, q^-, \mathbf{q}], \tag{2}$$

We also employ a number of energy rations following [23] $\omega \approx xE^+/2 = xP^+/2$, $\omega_0 \equiv \mathbf{k}^2/2\omega$, $\omega_i \equiv (\mathbf{k} - \mathbf{q}_i)^2/2\omega$, $\omega_{(ij)} \equiv (\mathbf{k} - \mathbf{q}_i - \mathbf{q}_j)^2/2\omega$, and $\tilde{\omega}_m \equiv \left(m_g^2 + M^2x^2\right)/2\omega$.

We make the following assumptions: 1) the eikonal, or high energy, approximation, for which E^+ is the largest energy scale of the problem; 2) the soft (radiation) approximation $x \ll 1$; 3) collinearity, $k^+ \gg k^-$; 4) that the impact parameter varies over a large transverse area; and, most crucially for the present work, 5) the large formation time assumption $\omega_i \ll \mu_i$, where $\mu_i^2 \equiv \mu^2 + \mathbf{q}_i^2$.

We have re-evaluated the 10 $N = 1$ in opacity diagrams without the large separation distance $\Delta z \gg 1/\mu$ assumption and present the results here.

IOP Conf. Series: Journal of Physics: Conf. Series **889** (2017) 012008 doi:10.1088/1742-6596/889/1/012008

3. Calculation and Results

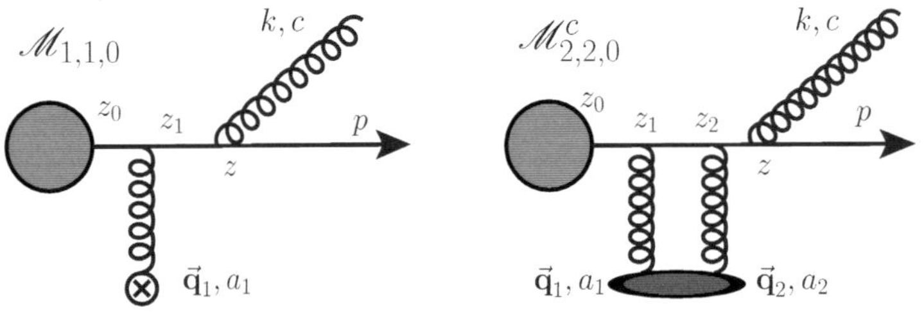

Figure 1: $\mathscr{M}_{1,1,0}$ (left panel) and $\mathscr{M}^c_{2,2,0}$ (right panel) are the only two diagrams that have non-zero short separation distance corrections in the large formation time limit. $\mathscr{M}^c_{2,2,0}$ is the double Born contact diagram, corresponding to the second term in the Dyson series in which two gluons are exchanged with the single scattering center.

The large separation distance assumption allowed the original DGLV calculation to neglect terms that are proportional to $\exp(-\mu\Delta z)$, which we have retained in our implementation of the relaxation of the large separation distance assumption. Nevertheless, an immense simplification occurs due to the large radiated gluon formation time approximation $\omega_i \ll \mu_i$ in that all but 2 of the 10 diagrams' 18 new small distance correction pole contributions are suppressed under the large formation time assumption. Fig. 1 shows the two diagrams that contribute non-zero small separation distance, large formation time terms to the result.

The full result for these two amplitudes under our approximation scheme is then

$$\begin{aligned}\mathscr{M}_{1,1,0} \approx & -J(p)e^{ipx_0}2gT_{a_1}ca_1 \int \frac{d^2\mathbf{q}_1}{(2\pi)^2} v(0,\mathbf{q}_1)e^{-i\mathbf{q}_1\cdot\mathbf{b}_1} \\ & \times \frac{\mathbf{k}\cdot\boldsymbol{\epsilon}}{\mathbf{k}^2+m_g^2+x^2M^2}\left[e^{i(\omega_0+\tilde{\omega}_m)(z_1-z_0)} - \frac{1}{2}e^{-\mu_1(z_1-z_0)}\right]\end{aligned} \tag{3}$$

$$\begin{aligned}\mathscr{M}^c_{2,2,0} \approx & \, J(p)e^{i(p+k)x_0} \int \frac{d^2\mathbf{q}_1}{(2\pi)^2} \int \frac{d^2\mathbf{q}_2}{(2\pi)^2} e^{-i(\mathbf{q}_1+\mathbf{q}_2)\cdot\mathbf{b}_1} \\ & \times igT_{a_2}T_{a_1}ca_2a_1 v(0,\mathbf{q}_1)v(0,\mathbf{q}_1)\frac{\mathbf{k}\cdot\boldsymbol{\epsilon}}{\mathbf{k}^2+m_g^2+x^2M^2} \\ & \times \left[e^{i(\omega_0+\tilde{\omega}_m)(z_1-z_0)} + e^{-\mu_1(z_1-z_0)}\left(1-\frac{\mu_1 e^{-\mu_2(z_1-z_0)}}{2(\mu_1+\mu_2)}\right)\right].\end{aligned} \tag{4}$$

It should be noted that the large formation time assumption allows one to neglect terms in both the original DGLV and the current short separation distance correction. The double differential single inclusive gluon emission distribution is given by [23]

$$d^3N_g^{(1)}d^3N_J = \frac{d^3\vec{\mathbf{p}}}{(2\pi)^3 2p^0}\frac{d^3\vec{\mathbf{k}}}{(2\pi)^3 2\omega}\left(\frac{1}{d_T}\operatorname{Tr}\langle|\mathscr{M}_1|^2\rangle + \frac{2}{d_T}\Re\operatorname{Tr}\langle\mathscr{M}_0^*\mathscr{M}_2\rangle\right). \tag{5}$$

Our main analytic result is then the $N=1$ first order in opacity small distance generalization of the DGLV induced energy loss of a high-p_T parton in a QGP:

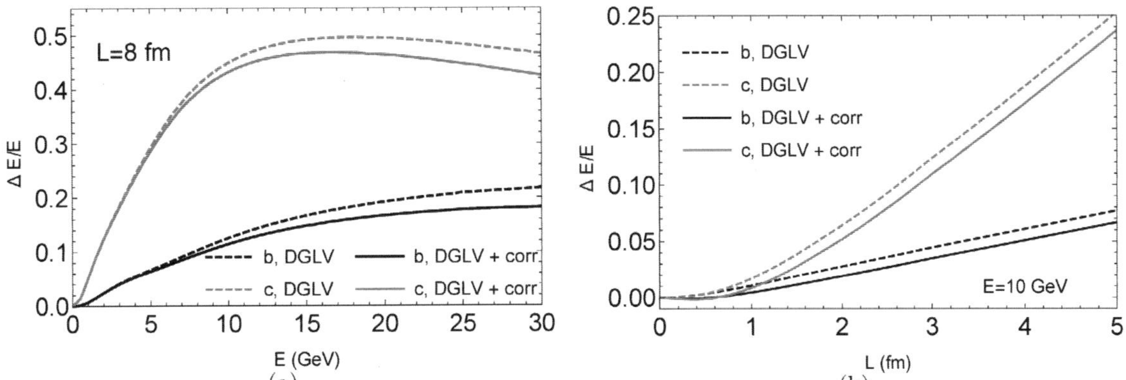

(a) (b)

Figure 2: Fractional energy loss of charm and bottom quarks in a QGP with $\mu = 0.5$ GeV and $\lambda_{mfp} = 1$ fm for (a) fixed path length $L = 8$ fm, and (b) fixed energy $E = 10$ GeV. In the figures, "DGLV" dashed curves are computed from the original $N = 1$ in opacity large separation distance DGLV formula while "DGLV + corr" solid lines are from our all separation distance generalization of the $N = 1$ DGLV result, equation (6).

$$\Delta E^{(1)}_{ind} = \frac{C_R\alpha_s LE}{\pi\lambda_g}\int dx\int\frac{d^2\mathbf{q}_1}{\pi}\frac{\mu^2}{(\mu^2+\mathbf{q}_1^2)^2}\int\frac{d^2\mathbf{k}}{\pi}\int d\Delta z\bar{\rho}(\Delta z)\Bigg[-\frac{2\left(1-\cos\{(\omega_1+\tilde{\omega}_m)\Delta z\}\right)}{(\mathbf{k}-\mathbf{q}_1)^2+m_g^2+x^2M^2}$$
$$\times\left(\frac{(\mathbf{k}-\mathbf{q}_1)\cdot\mathbf{k}}{\mathbf{k}^2+m_g^2+x^2M^2}-\frac{(\mathbf{k}-\mathbf{q}_1)^2}{(\mathbf{k}-\mathbf{q}_1)^2+m_g^2+x^2M^2}\right)$$
$$+\frac{1}{2}e^{-\mu_1\Delta z}\Bigg\{\left(\frac{\mathbf{k}}{\mathbf{k}^2+m_g^2+x^2M^2}\right)^2\left(1-\frac{2C_R}{C_A}\right)\left(1-\cos\{(\omega_0-\tilde{\omega}_m)\Delta z\}\right)$$
$$+\frac{\mathbf{k}\cdot(\mathbf{k}-\mathbf{q}_1)}{(\mathbf{k}^2+m_g^2+x^2M^2)\left((\mathbf{k}-\mathbf{q}_1)^2+m_g^2+x^2M^2\right)}\left(\cos\{(\omega_0-\tilde{\omega}_m)\Delta z\}-\cos\{(\omega_0-\omega_1)\Delta z\}\right)\Bigg\}\Bigg]. \tag{6}$$

The DGLV result appears in the first two lines of equation (6), while the last two are the short separation distance correction. The correction term behaves as expected by vanishing for both large and vanishing separation distances. However, the correction term also has a number of unexpected features: The correction term contains a color breaking term (proportional to $2C_R/C_A$) and dominates the original DGLV at high energies. This last feature is most apparent in the numerical analysis.

The numerical investigation of equation (6) is performed following [23] and results in figures 2a, 2b, 3a and 3b. The numerical analysis used the following values: $\mu = 0.5$ GeV, $\lambda_{mfp} = 1$ fm, $C_R = 4/3$, $C_A = 3$, $\alpha_s = 0.3$, $m_{charm} \equiv m_c = 1.3$ GeV and $m_{bottom} \equiv m_b = 4.75$ GeV, and the QCD analogue of the Ter-Mikayelian plasmon effect was taken into account by setting $m_{gluon} \equiv m_g = \mu/\sqrt{2}$. As in [24], kinematic upper limits were used for the momentum integrals such that $0 \le k \le 2x(1-x)E$ and $0 \le q \le \sqrt{3E\mu}$, due to finite kinematics. This choice of k_{max} guarantees that the final momentum of the parent parton is collinear to the initial momentum of the parent parton and that the momentum of the emitted gluon is collinear to the momentum of the parent parton. The fraction of momentum carried away by the radiated gluon, x was integrated over from 0 to 1. The distribution of scattering centers was assumed to be exponential in order to account for the rapidly expanding medium, $\bar{\rho}(z) = 2\exp(-2\Delta z/L)/L$.

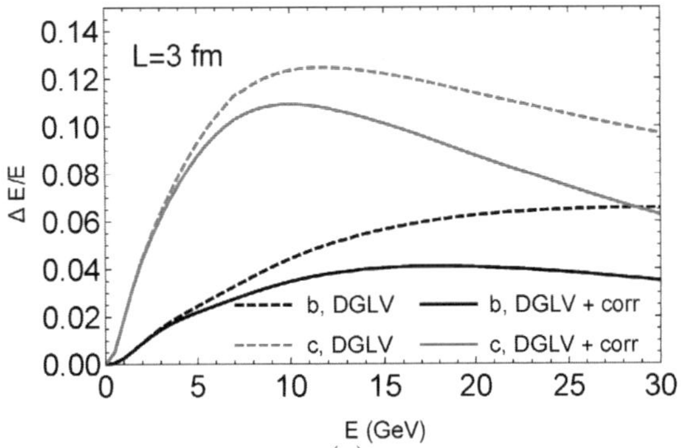

(a)

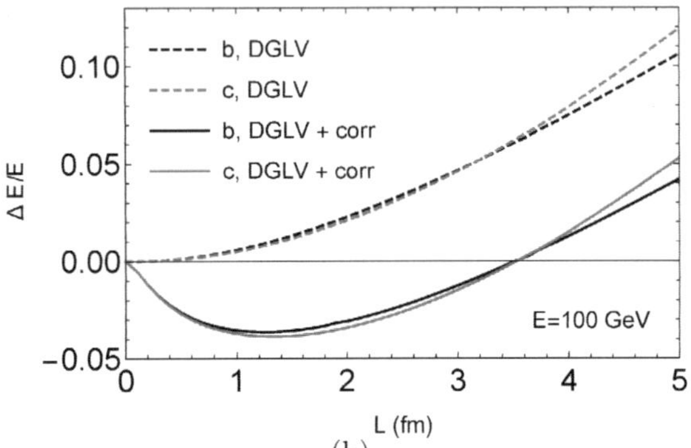

(b)

Figure 3: Fractional energy loss of charm and bottom quarks in a QGP with $\mu = 0.5$ GeV and $\lambda_{mfp} = 1$ fm for (a) fixed path length $L = 3$ fm, and (b) fixed energy $E = 100$ GeV. In the figures, "DGLV" dashed curves are computed from the original $N = 1$ in opacity large separation distance DGLV formula while "DGLV + corr" solid lines are from our all separation distance generalization of the $N = 1$ DGLV result, equation (6).

Fig. 2a shows the fractional energy loss of charm and bottom quarks of varying energy propagating through an 8 fm long static QGP brick. Clearly visible is that the small distance correction term is generally an energy *gain* due to the sign of the color triviality breaking term. Furthermore, relative to the large distance DGLV result, the size of the correction increases with energy.

Fig. 2b shows the fractional energy loss of charm and bottom quarks of energy $E = 10$ GeV for path lengths up to 5 fm. It is apparent from Fig. 2b that the small separation distance correction has an appreciable effect even for large path lengths. This feature is well understood as being due to the integration over all separation distances between the production point and the scattering position, but this effect diminishes with path length, as it must.

Fig. 3a illustrates the effect on small systems by showing the fractional energy loss of charm and bottom quarks through a, much smaller, 3 fm thick static QGP brick. Here, the striking effect of the energy of the parent parton on its energy loss in a small system is clearly visible. To illustrate this point further, Fig. 3b shows the fractional energy loss of 100 GeV charm and bottom quarks propagating up to 5 fm through a QGP. Clearly, the small distance "correction" term to the DGLV result dominates, even out to path lengths of ~ 3 fm, for high energies.

4. Conclusions

It is difficult to reconcile the significant *energy gain* at high energies with either experimental suppression of charged particles in central AA collisions at LHC [13–15] or an intuitive understanding of the assumption scheme that is employed. As such, a more detailed interrogation of the validity of the large formation time assumption in pQCD-based energy loss calculations is necessary.

5. Acknowledgments

The authors wish to thank the SA-CERN Collaboration, the South African National Research Foundation (NRF), and the National Institute for Theoretical Physics (NITheP) for their generous support. The authors also wish to thank CERN for the hospitality extended to them during the completion of part of this work. Finally, the authors wish to thank Miklos Gyulassy, Ulrich Heinz, Sangyong Jeon, and Carlos Salgado for valuable discussions.

References

[1] V. Khachatryan *et al.*, "Evidence for Collective Multiparticle Correlations in p-Pb Collisions," *Phys. Rev. Lett.*, vol. 115, no. 1, p. 012301, 2015.
[2] B. B. Abelev *et al.*, "Multiplicity Dependence of Pion, Kaon, Proton and Lambda Production in p-Pb Collisions at $\sqrt{s_{NN}}$ = 5.02 TeV," *Phys. Lett.*, vol. B728, pp. 25–38, 2014.
[3] CMS Collaboration, "Multiplicity and rapidity dependence of strange hadrons spectra in pp, pPb, and PbPb collisions at LHC energies," 2015.
[4] J. Adam *et al.*, "Centrality dependence of inclusive J/ production in p-Pb collisions at $\sqrt{s_{\rm NN}}$ = 5.02 TeV," *JHEP*, vol. 11, p. 127, 2015.
[5] A. Adare *et al.*, "Centrality-dependent modification of jet-production rates in deuteron-gold collisions at $\sqrt{s_{NN}}$=200 GeV," 2015.
[6] G. Aad *et al.*, "Centrality and rapidity dependence of inclusive jet production in $\sqrt{s_{\rm NN}}$ = 5.02 TeV proton-lead collisions with the ATLAS detector," *Phys. Lett.*, vol. B748, pp. 392–413, 2015.
[7] N. Armesto, M. Cacciari, A. Dainese, C. A. Salgado, and U. A. Wiedemann, "How sensitive are high-p(T) electron spectra at RHIC to heavy quark energy loss?," *Phys. Lett.*, vol. B637, pp. 362–366, 2006.
[8] W. A. Horowitz, "Heavy Quark Production and Energy Loss," *Nucl. Phys.*, vol. A904-905, 2013.
[9] K. M. Burke *et al.*, "Extracting the jet transport coefficient from jet quenching in high-energy heavy-ion collisions," *Phys. Rev.*, vol. C90, no. 1, p. 014909, 2014.
[10] M. Djordjevic, M. Djordjevic, and B. Blagojevic, "RHIC and LHC jet suppression in non-central collisions," *Phys. Lett.*, vol. B737, pp. 298–302, 2014.
[11] A. Adare *et al.*, "Azimuthal anisotropy of π^0 and η mesons in Au + Au collisions at $\sqrt{s_{NN}}$ = 200 GeV," *Phys. Rev.*, vol. C88, no. 6, p. 064910, 2013.
[12] B. B. Abelev *et al.*, "Neutral pion production at midrapidity in pp and Pb-Pb collisions at $\sqrt{s_{\rm NN}}$ = 2.76 TeV," *Eur. Phys. J.*, vol. C74, no. 10, p. 3108, 2014.
[13] S. Chatrchyan *et al.*, "Study of high-pT charged particle suppression in PbPb compared to *pp* collisions at $\sqrt{s_{NN}}$ = 2.76 TeV," *Eur. Phys. J.*, vol. C72, p. 1945, 2012.
[14] B. Abelev *et al.*, "Centrality Dependence of Charged Particle Production at Large Transverse Momentum in Pb–Pb Collisions at $\sqrt{s_{\rm NN}}$ = 2.76 TeV," *Phys. Lett.*, vol. B720, pp. 52–62, 2013.
[15] G. Aad *et al.*, "Measurement of charged-particle spectra in Pb+Pb collisions at $\sqrt{s_{\rm NN}}$ = 2.76 TeV with the ATLAS detector at the LHC," *JHEP*, vol. 09, p. 050, 2015.
[16] B. I. Abelev *et al.*, "Transverse momentum and centrality dependence of high-p_T non-photonic electron suppression in Au+Au collisions at $\sqrt{s_{NN}}$ = 200 GeV," *Phys. Rev. Lett.*, vol. 98, p. 192301, 2007. [Erratum: Phys. Rev. Lett.106,159902(2011)].
[17] S. Sakai, "Measurement of R_{AA} and ν_2 of electrons from heavy-flavour decays in Pb-Pb collisions at $\sqrt{s_{NN}}$ = 2.76 TeV with ALICE," *Nucl. Phys.*, vol. A904-905, pp. 661c–664c, 2013.
[18] A. Adare *et al.*, "Single electron yields from semileptonic charm and bottom hadron decays in Au+Au collisions at $\sqrt{s_{NN}}$ = 200 GeV," 2015.
[19] J. Adam *et al.*, "Transverse momentum dependence of D-meson production in Pb-Pb collisions at $\sqrt{s_{\rm NN}}$ = 2.76 TeV," 2015.
[20] S. Chatrchyan *et al.*, "Suppression of non-prompt J/ψ, prompt J/ψ, and Y(1S) in PbPb collisions at $\sqrt{s_{NN}}$ = 2.76 TeV," *JHEP*, vol. 05, p. 063, 2012.
[21] U. A. Wiedemann, "Jet Quenching in Heavy Ion Collisions," pp. 521–562, 2010. [Landolt-Bornstein23,521(2010)].
[22] M. Gyulassy, P. Levai, and I. Vitev, "Reaction operator approach to nonAbelian energy loss," *Nucl. Phys.*, vol. B594, pp. 371–419, 2001.
[23] M. Djordjevic and M. Gyulassy, "Heavy quark radiative energy loss in QCD matter," *Nucl. Phys.*, vol. A733, pp. 265–298, 2004.
[24] S. Wicks, W. Horowitz, M. Djordjevic, and M. Gyulassy, "Elastic, inelastic, and path length fluctuations in jet tomography," *Nucl. Phys.*, vol. A784, pp. 426–442, 2007.
[25] W. A. Horowitz and B. A. Cole, "Systematic theoretical uncertainties in jet quenching due to gluon kinematics," *Phys. Rev.*, vol. C81, p. 024909, 2010.
[26] N. Armesto *et al.*, "Comparison of Jet Quenching Formalisms for a Quark-Gluon Plasma 'Brick'," *Phys. Rev.*, vol. C86, p. 064904, 2012.
[27] I. Kolbé and W. A. Horowitz, "Short path length pQCD corrections to energy loss in the quark gluon plasma," arXiv:1509.06122.
[28] M. Gyulassy and X.-n. Wang, "Multiple collisions and induced gluon Bremsstrahlung in QCD," *Nucl. Phys.*, vol. B420, pp. 583–614, 1994.

IOP Conf. Series: Journal of Physics: Conf. Series **889** (2017) 012009 doi:10.1088/1742-6596/889/1/012009

High fluence neutron radiation of plastic scintillators for the TileCal of the ATLAS detector.

J E Mdhluli[a,b,1], Yu I Davydov[c], V Baranov[c], S Mthembu[a], R Erasmus[a], H Jivan[a], N Khanye[a], H Tlou[a], B Tjale[a], J Starchenko[d], O Solovyanov[d], B Mellado[a] and E Sideras-Haddad[a,b]

[a] School of Physics, University of the Witwatersrand, Johannesburg, Wits 2050, South Africa.
[b] DST-NRF Centre of Excellence in Strong Materials.
[c] Joint Institute for Nuclear Research, Dubna, Russia.
[d] Institute for High Energy Physics, Protvino 142281, Russia

E-mail: [1]joyemmie@gmail.com

Abstract. We report on structural and optical properties of neutron irradiated plastic scintillators. These scintillators were subjected to a neutron beam with wide energy range of up to $10MeV$ and a neutron flux range of $1.2 \times 10^{12} - 9.4 \times 10^{12} n/cm^2$ using the IBR-2 pulsed reactor at the Joint Institute for Nuclear Research in Dubna. A study between polyvinyl toluene based commercial scintillators EJ200, EJ208 and EJ260 as well as polystyrene based scintillator from Kharkov is conducted. Light transmission, Raman spectroscopy, fluorescence spectroscopy and light yield testing was performed to characterize the damage induced in the samples. Preliminary results from the tests performed indicate no change in the optical and structural properties of the scintillators. The polystyrene based scintillators were further subjected to a higher neutron flux range of $3.8 \times 10^{12} - 1.8 \times 10^{14} n/cm^2$ using the IBR-2 pulsed reactor.

1. Introduction

The ATLAS detector is one of the four interaction points on the Large Hadron Collider (LHC) ring together with CMS, ALICE and LHCb detectors. It is used in particle physics experiments that are involved in the search for new particles through high energy proton-proton collisions at the LHC of CERN. The detector consists of six different detecting subsystems that are arranged in layers around the collision point to record the path, momentum and energy of the particles in order for the particles to be individually identified [1]. The tile calorimeter is a hadronic calorimeter within the calorimetry subsystem of the detector responsible for detecting hadrons, taus, and jets of quarks and gluons. It consists of a central barrel and 2 extended barrels with each barrel containing 64 modules that are made of a matrix of steel plates and plastic scintillator tiles sandwiched between them. The incoming jets are converted into a "shower" of

IOP Conf. Series: Journal of Physics: Conf. Series **889** (2017) 012009 doi:10.1088/1742-6596/889/1/012009

particles by the steel plates. Thereafter, the plastic scintillator tiles absorb the energy of the particles and fluorescence to emit light. The light from the scintillators is passed through wavelength shifting (WLS) optical fibres and is detected by photomultiplier tubes. The signal is further processed using readout electronics in order to digitize the data for further analysis [2] [3].

Further plastic scintillators are found between the central and extended barrels, this area is known as the gap region. During the first run of data taking, the scintillators in the gap region were exposed to a radiation environment of up to 10 kGy/year. It is predicted that during the high luminosity (HL)-LHC run time, the scintillators in the Gap region will sustain a significantly large amount of radiation damage and will require replacement. This prediction has led to the comparative study of proton induced radiation damage on plastic scintillators conducted by H Jivan [2] [4], C Pelwan [5] and S Liao [6] where it was concluded that scintillators with larger Stokes shift, i.e. EJ260 and EJ208, exhibit the most radiation hardness.

The comparative study has been extended to a neutron irradiation and damage assessment study on plastic scintillators. Neutron interaction with matter makes the study more interesting since unlike with proton irradiation where the interaction with the plastics is through direct ionization, with neutron irradiation the interaction will be through indirect ionization. Neutron bombardment on materials creates a collision cascade within the material that results to point defects and dislocations. The collisions cause a massive transfer of kinetic energy to the lattice atom that has been displaced from its lattice site, becoming what is known as the primary knock-on atom (PKA). The knock-on atoms then lose energy with each collision and that energy therefore ionizes the material [7].

2. Experimental Procedure

Four plastic scintillator grades were under study, three of which were obtained from ELJEN Technologies and one from ISMA, Kharkov.The plastic scintillator grades under study are the EJ200, EJ208, EJ260 and the UPS-932A type. The three plastic scintillators obtained from ELJEN technologies are composed of a polyvinyl toluene base and 3% added organic fluors [1], the fourth plastic scintillator grade is the UPS-932A and is composed of polysterene base.

Several samples of each plastic scintillator grade were cut and polished to dimensions 20 mm by 20 mm by 6 mm thickness at the Dzhelepov Laboratory of Nuclear Problems (DLNP) at JINR. Special sample holders were made to accommodate the samples for irradiation. Channel 3 of the IBR-2 pulsed reactor at the Frank Laboratory of Neutron Physics (FLNP) at JINR was used to irradiate the samples [8]. The samples were subjected to irradiation with a beam of fast neutrons with energies in the wide range of up to 10 MeV for 337 hours. The reactor was operating at an average power of 1875 kW. The samples were placed at three different positions from the reactor core to expose them to different neutron fluences in order to achieve various doses. The neutron

Table 1. Neutron flux density, neutron fluences and doses at the various positions.

Sample position	**Flux density** $(n/cm^2/s)$	**Fluence** (n/cm^2)
1	1×10^6	1.2×10^{12}
2	3.6×10^6	3.6×10^{12}
3	7.7×10^6	9.4×10^{12}

flux density ranged between 1×10^6 - $7.7 \times 10^6 n/cm^2/s$.

The Monte Carlo N-Particle (MCNP) 5 [9] code was used to simulate neutron transport through plastic scintillators and to determine the dose rate. The MCNP code is used to simulate neutron, photon and electron or coupled neutron/photon/electron transport. It is in 3-dimensions and is capable of tracking up to 34 particles and 4 light ions. It takes into account the absorption and the moderation of the neutrons as they travel through matter. Table 1 below shows the doses for various neutron flux density and neutron fluences. The Varian Carry 500 spectrophotometer was used to characterize the optical properties of the irradiated samples due to the damage of the neutron irradiation. The light transmission of the samples was measured relative to the transmission in air over a laser wavelength range between 200-800 nm.

The structural damage the scintillators may have undergone after irradiation were measured using the Raman spectroscopy technique. The Horiba Jobin-Yvon Raman spectrograph consists of an Argon laser used to provide a 515 nm excitation wavelength. Raman spectra were obtained for the un-irradiated control samples as well as on the irradiated samples to gauge the changes in the samples.

Fluorescence testing was measured with the aid of LabrRAM HR Raman spectroscopy setup. A laser of wavelength of 229 nm with a power ranging from $3 \sim 5$ mW was employed to provide enough energy for molecular excitation, which results in prompt light emission through luminescence. The laser is guided through a series of mirrors and optics in the setup, and is incident on the sample through the microscope aperture. As the sample fluoresces, the light emitted in the direction back up the aperture is collected by a detector and a differential wavelength spectrum is obtained.

Light yield testing was conducted at CERN. The samples were excited with a Sr-90 source that undergoes beta decay. As the sample flouresces to emit light, the light is passed through four WLS optical fibres and detected by a PMT. The signal from the PMT is further processed through electronics and is digitized in the computer. For better contact between the sample and the fibres, profiles were placed along the sides of the sample to ensure the stability of the fibres.

3. Results and Discussion

Figure 1 shows the light transmission spectra of the EJ200. It is observed that at a wavelength of 400 nm the absorptive edge falls away completely and that the overall transmission of the grade remains unchanged even after irradiation. The transmission

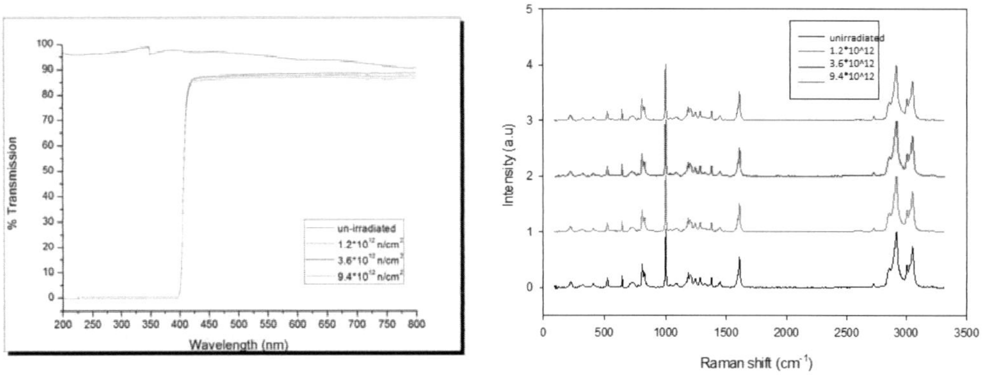

Figure 1. Left: Light Transmission spectra for the EJ200 samples. Right: Raman spectra for the EJ200 samples.

loss is observed at wavelength 450 nm as this corresponds to the peak absorption wavelength of the wavelength shifting optical fibres coupled with these scintillators within the Tile Calorimeter. Even at this wavelength, there is still no change observed in the samples with increasing doses. The other scintillator grades also behave in the same manner.

Raman spectra were obtained for the various irradiated samples as well as for un-irradiated control samples. The irradiated samples maintained their structure with no additional peaks being formed and no damage was observed. Figure 1 also shows the Raman spectra of the EJ200 sample for the irradiated and un-irradiated samples with the highest neutron fluence at the top and the un-irradiated at the bottom. The other scintillator grades also behave in the same manner.

Figure 2 shows the fluorescence spectra of the EJ200 sample. Several fluorescence peak features are observed in the spectra. Peaks in the wavelength region of 300-375 nm correlate to fluorescence of the PVT/PS base. The "two-peak" feature is observed since the fluorescence is predominantly from the benzene ring structure. In the wavelength region of 375-500 nm, the fluorescence correlates to that of the fluor dopants. No change in fluorescence is observed in the spectra for the irradiated and un-irradiated samples. The other scintillator grades also behave in the same manner.

The light yield of the scintillators is also shown in figure 2. The light yield of the samples was measured by testing their response to a Sr-90 source that undergoes beta decay. This setup was aimed at mimicking what happens within the detector when the scintillators are measuring the energy of the particles. The EJ260 samples showed the lowest light yield. This can be a result of the maximum emission (490 nm) of the scintillator grade being greater than the maximum absorption (435 nm) of the wavelength shifting optical fibre coupled to it. The EJ200 and the EJ208 samples have a maximum emission of 425 nm and 435 nm respectively [10], allowing for the maximum amount of light emitted by the scintillators to be absorbed by the WLS optical fibres. From the graph below, we observe no change for all the scintillator

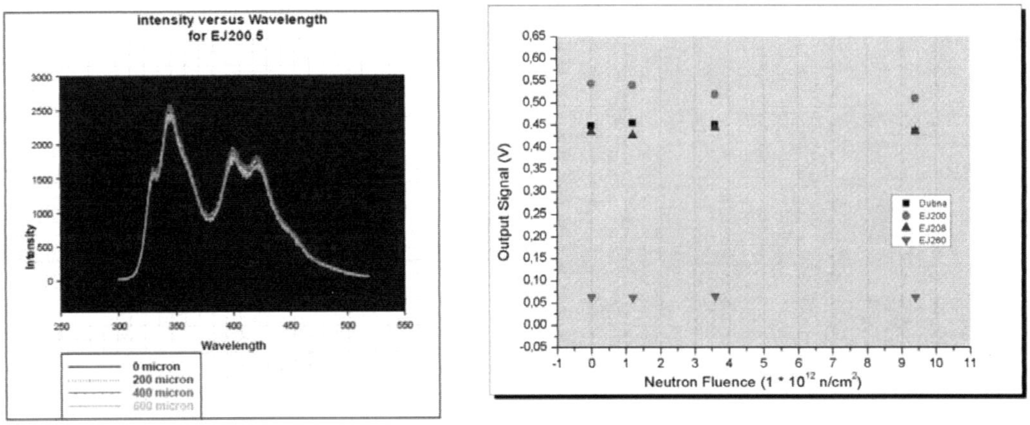

Figure 2. Left: Fluorescence spectra for the EJ200 sample. Right: Light yield output signals versus neutron fluence.

grades with irradiation.

4. Conclusion

The radiation damage the plastics were exposed to is very low relative to that within the ATLAS detector. Thus, as observed in the results for the various tests, there was no change in the structural or optical properties of the plastics for the first set of samples that were irradiated during the April 2016 run. The study needs to be extended to higher neutron fluences.

5. Further work

During the autumn 2016 run of the IBR-2 reactor, UPS-923A scintillators were irradiated at higher neutron fluences range of 3.8×10^{-12} - 1.8×10^{14} n/cm^2. Raman Spectroscopy, Light yield measurements and Light transmission testing were performed on the irradiated samples. Light yield measurements were done at the Dzhelepov Laboratory of Nuclear Problems at the Joint Institute for Nuclear Research. The light yield decreased by $\sim 28\%$ after the neutron irradiation of 1.8×10^{14} n/cm^2. A decrease in the light transmission was also observed from the light transmission spectrum. No evident structural changes were observed after the irradiation from the Raman spectroscopy measurements. No additional peaks were formed after irradiation. A more detailed analysis of these results is discussed in a proceedings written by Skhathisomusa Mthembu in this conference series.

IOP Conf. Series: Journal of Physics: Conf. Series **889** (2017) 012009 doi:10.1088/1742-6596/889/1/012009

6. Acknowledgements

We would like to acknowledge the Frank Laboratory of Neutron Physics for making the IBR-2 reactor available for experiments and the staff involved.

References

[1] ATLAS:CERN [Online] Available: *https://home.cern/* [Accessed 24 08 2016]
[2] Jivan H, Erasmus R, Mellado B, Peters G, Sekonya K, Sideras-Haddad E 2014 Radiation hardness of plastic scintillators for the Tile Calorimeter of the ATLAS detector (Submitted to the SAIP conference)
[3] Succurro A 2012 *Physcis Procedia* **37** 229-237
[4] Jivan H, Mellado B, Sideras-Haddad E, Erasmus R, Liao S, Madhuku M, Peters G, Solvyanov O 2015 *Journal of Physics conference series* **645** 012019
[5] Pelwan C, Jivan H, Joubert D, Keartland J, Liao S, Peters G, Sideras-Haddad E 2015 *Journal of Physics conference series* **10** 645:012023
[6] Liao S, Erasmus R, Jivan H, Pelwan C, Peters G, Sideras-Haddad E 2015 *Journal of Physics conference series* **10** 645:012021
[7] Bisanti P 1986 *Magnetic properties of matter* (Turin:World Scientific) ed Borsa F, Tognetti V p 385-408
[8] Bulavin M. 2015 Irradiation facility at the IBR-2 reactor for investigation of material radiation hardness *NIM* **B343** p26 - 29.
[9] Monte Carlo Group: Los Alamos National Laboratory [Online] Available: *https://mcnp.lanl.gov/* [Accessed 24 11 2016]
[10] ELJEN Technology 2013 Products: Plastic Scintilators [Online] Available: [Accessed 26 08 2016]

IOP Conf. Series: Journal of Physics: Conf. Series **889** (2017) 012010 doi:10.1088/1742-6596/889/1/012010

The performance of diphoton primary vertex reconstruction methods in H→ $\gamma\gamma$+Met channel of ATLAS experiment

K G Tomiwa[1*]

[1] School of Physics, University of the Witwatersrand, Johannesburg 2050, South Africa

E-mail: kehinde.gbenga.tomiwa@cern.ch

Abstract. The search for new physics in the H→ $\gamma\gamma$+met relies on how well the missing transverse energy is reconstructed. The Met algorithm used by the ATLAS experiment in turns uses input variables like photon and jets which depend on the reconstruction of the primary vertex. This document presents the performance of di-photon vertex reconstruction algorithms (hardest vertex method and Neural Network method). Comparing the performance of these algorithms for the nominal Standard Model sample and the Beyond Standard Model sample, we see the overall performance of the Neural Network method of primary vertex selection performed better than the Hardest vertex method.

1. Introduction

The reconstruction of the physics process ($H \rightarrow \gamma\gamma$) of this analysis requires precise measurement of the primary vertex (PV) [1]. Reconstructed interaction vertices are used by algorithm [2] that distinguishes measurements from the hard-scattering collision (signal) and additional collision ("pileup"). The Jet-Vertex-Tagger (JVT) also relies on precise reconstruction of the interaction vertices [3]. The common approach used for vertex interaction point is to identify the primary vertex with the highest scalar sum of transverse momenta ($\sum p_T$) of tracks (choice of most ATLAS analyses), this method is sub-optimal with photon because photons often do not leave tracks in the inner detector. A more efficient way of reconstructing photon vertex interaction point is to take advantage of the longitudinal segmentation of the electromagnetic calorimeter (this method is often referred to as "pointing") [1]. Interaction vertices are reconstructed using signatures from the inner detector (see Sec 2) In preparation for Run-II a Neural network algorithm (NN algorithm as soon known as pointing method) was developed, this algorithm combines the two approaches to a framework that discriminates between the hard-scatter vertex (signal) and vertices associated with additional collision in same bunch crossing(in-time pileup). $H \rightarrow \gamma\gamma$ analysis.

2. ATLAS Tracking System

The ATLAS experiment uses the Inner detector (ID) system (see Figure 1) as the major tracking system which is made up of a silicon pixel detector, a silicon microstrip detector (SCT), the straw tubes of the transition radiation tracker (TRT), and the insertable B-layer (IBL) recently added

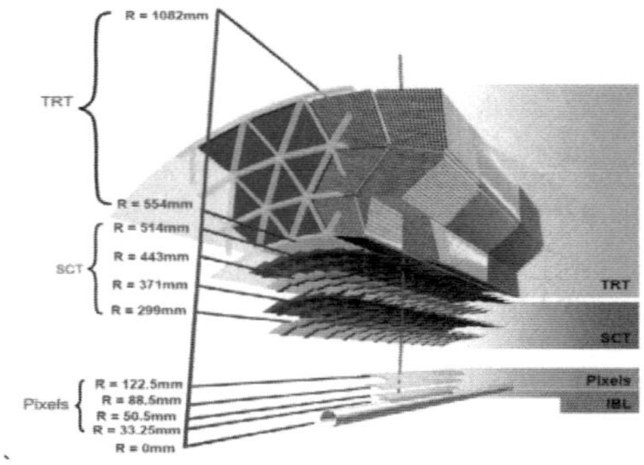

Figure 1. Sketch of the ATLAS inner detector showing all its components, including the new insertable B-layer (IBL). The distances to the interaction point are also shown.

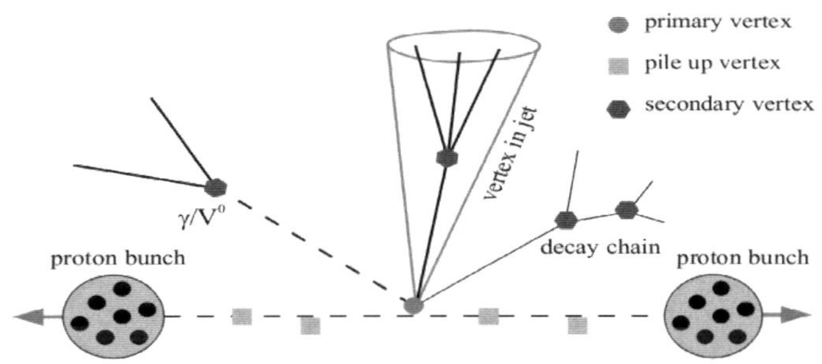

Figure 2. Vertex topologies in ATLAS primary and pile up vertices from conversions and long-lived particles, vertices in jets and vertices from chain decay.

to ATLAS detector. The ID is designed to measure the trajectories and momentum of charged particle in the region of $\eta < 2.5$[1] [4].

3. Vertex topologies in ATLAS Detector

Several different vertex topologies arise from the collision of bunches of proton in the LHC at the center-of-mass energy of 13 TeV as shown in Fig 2. In a typical collision event, several primary vertices along the beam and decay of long-lived particles, photon conversion, vertices in jets and vertices from decay chains are produced. Hence, the reconstruction of these vertices from processes and distinguished by their different topology are very important. The photon conversion vertices is particularly important to the analysis in $H \rightarrow \gamma\gamma$ decay channel [2].

There are different types of photon conversion considered for the $H \rightarrow \gamma\gamma$ analysis [5]. Double track photon conversions are photons that convert early in the inner detector with separated trajectories, for the photon which conversion occurs at higher radius or if reconstruction of two separate tracks fails the photon is referred to as single track conversion. The TRT standalone (TRTSA) conversion are photons which converted at the TRT with imprecise z

[1] ATLAS coordinated with respect to the interaction point (IP) is such that: the z-axis is along the beam pipe. The x-axis points from the IP to the center of the LHC ring and the y-axis points upward. The ATLAS transverse plan uses Cylindrical coordinates (r,ϕ), ϕ is azimuthal angle around the z-axis, the pseudorapidity (η) is defined as $-\ln tan(\theta/2)$, where θ is the polar angle. Angular distance is measured in the unit od $\Delta = \sqrt{\Delta^2 + \phi^2}$

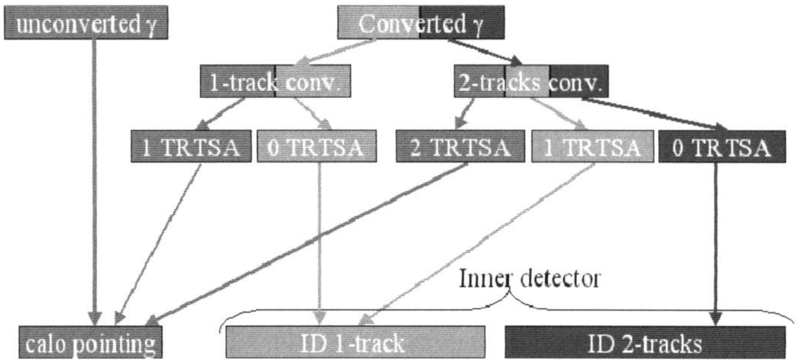

Figure 3. Strategy for the recostruction of the z primary vertex and $\gamma\gamma$.

and R measurement in the barrel and endcap respective. TRTSA photons are reconstructed using direction information from the electromagnetic calorimeter [2]. The global strategy and the method to be used as a function of the nature of photon is shown in Fig. 3

4. Selection of $\gamma\gamma$ production vertex using multivariate technique

The motivation for developing this techniques is the increase in pileup events due to high luminosity of the LHC collision, the increase in luminosity increases the number of interaction in the same bunch crossing. Different discriminant (artificial neural network, boosted decision trees, k-Nearest Neighbors and likelihood) algorithm were trained on simulated samples (sample containing pileup with 50 ns bunch spacing) containing Higgs bosons produced through gluon-gluon fusion and decay into two photons (H→ $\gamma\gamma$), the performance of the algorithms was cross-validated and evaluated using samples with 25 ns pileup. The algorithms were trained to distinguished:

- At least one photon converted into e^+e^- pairs and with track associated with the primary vertex.
- At least one converted photo with the track containing hit the in the silicon detector.
- Both photon unconverted or associated to track with a hit in the TRT.

The following inputs are used in the discriminants:

- $(z_{\text{common}} - z_{\text{vertex}})/\sigma_z$,)($z_{\text{vertex}}$ position of primary vertex) and z_{common} (intersection of the extrapolated photon trajectories from the calorimeter("pointing"))
- $\log_{10}(\sum pt)$, the scalar sum of transverse momenta of the tracks associated to the vertex.
- $\log_{10}(\sum pt^2)$, idem with p_T^2 variable transformed to ensure the convergence of minimizer.
- $\Delta\phi(\gamma\gamma, vertex)$: the angular difference between the diphoton system and system defined by the vector sum of the tracks associated to the vertex.

The neural network discriminant was chosen because it has the best efficiency [1]. Hence the neural network method will be refereed to as the pointing method in this document.

5. Sample and Event Selection

The event selection used for this studies is based on the $H \rightarrow \gamma\gamma$ nominal selection [3] with additional missing transverse momentum (MET) requirement, two photon selection with transverse momentum of the leading and sub-leading photon to be greater than 0.35 and 0.25 respectively. The heavy scalar signal sample is used for this analysis, the sample has process that models the production of Higgs boson through gluon-gluon fusion.

IOP Conf. Series: Journal of Physics: Conf. Series **889** (2017) 012010 doi:10.1088/1742-6596/889/1/012010

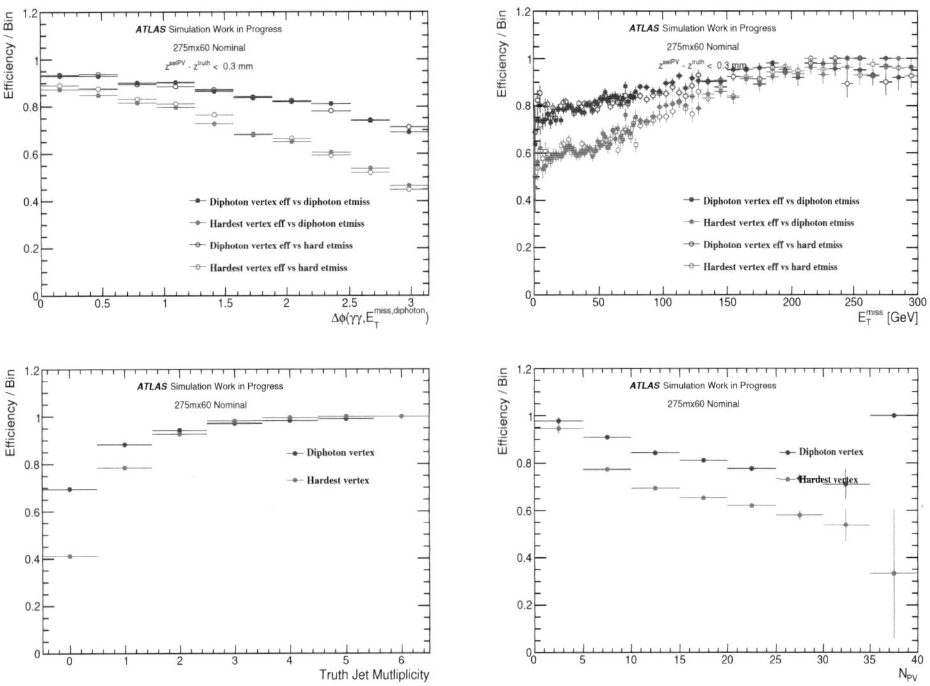

Figure 4. Efficiency for selecting the corrected primary vertex as a function of $\Delta\phi(\gamma\gamma, E_T^{miss,photon})$, E_T^{miss}, Truth Jet Multiplicity and number of primary vertex when considering the hardest vertex or neural network algorithm.

6. Result

The efficiency of identifying correct vertex is calculated for reconstructed vertex which matches the truth vertex with a z-direction of < 0.3 mm.The efficiency is calculated as the ratio of the number of events that pass this criteria to the total number of events ($\epsilon = \frac{N_{Evt}^{pass}}{N_{Evt}^{tot}}$). Figure 4 to 5 show the efficiency selecting correct vertex for photon pointing method and hardest vertex method as a function of Number of primary vertex, missing transverse energy (E_T^{miss}), truth photon leading momentum (Truth $p_T^{\gamma\gamma}$), angular distance between the diphoton system and the vector system of E_T^{miss}, jet multiplicity and sum squared of photon momentum $\sum pt^2$.

From Fig. 4 the efficiency of the photon pointing method and hardest vertex method decreases as a function of angular distance between the E_T^{miss} energy and diphoton system, a similar distribution is observed for the number of primary vertex. And from the E_T^{miss} plot, we see higher efficiency of photon pointing method for low E_T^{miss} ($E_T^{miss} < 100$(GeV)), efficiency are close at higher E_T^{miss}. Both methods have similar distribution for the Truth Jet multiplicity. From Fig. 5 the efficiencies increase with increase in the variables (Truth $p_T^{\gamma\gamma}$, $|E_T^{miss,photon} - p_T^{\gamma\gamma}|/p_T^{\gamma\gamma}$, sum of p_T and sum of p_T^2). The efficiencies become constants and comparable at high Truth $p_T^{\gamma\gamma}$, sum of p_T and sum of p_T^2 for both methods.

7. Conclusions

The search for Higgs in association with missing transverse energy relies on precise measurement of missing energy, the missing transverse algorithm uses physics objects like the diphoton system and jets as input. Diphoton systems are often contaminated by ambiguity in distinguishing the unconverted photon and electron likewise the jet system are affected by pileup jets. Photon

IOP Conf. Series: Journal of Physics: Conf. Series **889** (2017) 012010 doi:10.1088/1742-6596/889/1/012010

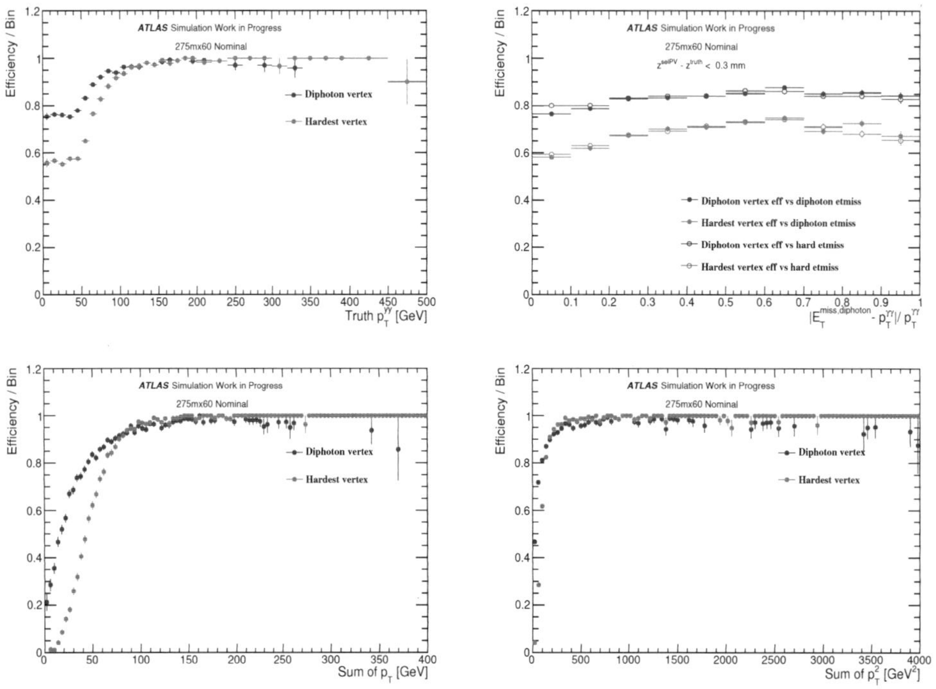

Figure 5. Efficiency for selecting the corrected primary vertex as a function of Truth $p_T^{\gamma\gamma}$, $|E_T^{miss,photon} - p_T^{\gamma\gamma}|/p_T^{\gamma\gamma}$, sum of p_T ans sum of p_T^2 when considering the hardest vertex or neural network algorithm.

isolation and jet vertex tagging methods are ways the ATLAS experiment uses to reduce this effect in most analysis. From Figure 4 to 5 we see the photon pointing method in selecting correct vertex outperforms the hardest vertex method. The photon pointing method has been recommended for most analysis in ATLAS experiment. And based on this study it is the most efficiency method or the search for Higgs boson in association with intermediate missing energy.

References

[1] Lenzi B and Delgove D 2015 Selection of the diphoton production vertex using multivariate techniques for $H \rightarrow \gamma\gamma$ and other analyses Tech. Rep. ATL-COM-PHYS-2015-1321 CERN Geneva URL `https://cds.cern.ch/record/2062239`

[2] Bouhova-Thacker E, Lichard P, Kostyukhin V, Liebig W, Limper M, Piacquadio G, Lichard P, Weiser C and Wildauer A 2009

[3] collaboration T A (ATLAS) 2016

[4] ATLAS-Collaboration 2015 Track Reconstruction Performance of the ATLAS Inner Detector at $\sqrt{s} = 13$ TeV Tech. Rep. ATL-PHYS-PUB-2015-018 CERN Geneva URL `https://cds.cern.ch/record/2037683`

[5] Aad G *et al.* (ATLAS) 2009 (*Preprint* `0901.0512`)

IOP Conf. Series: Journal of Physics: Conf. Series **889** (2017) 012011 doi:10.1088/1742-6596/889/1/012011

Performance of a resistive plate chamber equipped with a new prototype of amplified front-end electronics in the ALICE detector

Massimiliano Marchisone
iThemba LABS, National Research Foundation, Somerset West, South Africa
University of the Witwatersrand, Johannesburg, South Africa

E-mail: mmarchis@cern.ch

Abstract. ALICE is the LHC experiment dedicated to the study of heavy-ion collisions. At forward rapidity a muon spectrometer detects muons from low mass mesons, quarkonia ($c\bar{c}$ and $b\bar{b}$ mesons), open heavy-flavor hadrons (D and B mesons) as well as from weak bosons.

A muon selection based on transverse momentum is made by a trigger system composed of 72 Resistive Plate Chambers (RPCs). For the LHC Run 1 and the ongoing Run 2 the RPCs have been equipped with a non-amplified Front-End Electronics (FEE) called ADULT. However, in view of an increase in luminosity expected for Run 3 (foreseen to start in 2021) the possibility to use an amplified FEE has been explored in order to improve the counting rate limitation and to prevent the aging of the detector by reducing the charge per hit. A prototype of this new electronics (FEERIC) has been developed and tested first with cosmic rays before equipping one RPC in the ALICE cavern with it.

In this proceeding the most important performance indicators (such as efficiency, dark current, dark rate, cluster size, total charge and charge per hit) of the RPC equipped with this new FEE will be reviewed and compared to the others read out with ADULT.

1. Introduction

ALICE [1] is a multitasking experiment at the LHC at CERN, mainly designed to study the hot and dense matter produced in heavy-ion collisions. At forward rapidity[1] ($2.5 < y < 4$) muons are identified and tracked with a muon spectrometer [2], consisting of a set of absorbers, a dipole magnet, a muon tracking system and a muon trigger detector [3].

The muon trigger system consists of 72 Resistive Plate Chambers (RPCs) [4] arranged in two stations at a distance of 16 and 17 m from the interaction point. Each station is made of two detection planes with 18 RPCs each, perpendicular to the beam line. The chambers are 2 mm single gap RPC, with low resistivity ($\sim 10^9$ $\Omega\cdot$cm) [5] bakelite electrodes and are operated in the so-called "maxi-avalanche" mode [6], with a gas mixture consisting of 89.7% $C_2H_2F_4$, 10% C_4H_{10} and 0.3% SF_6. The gas relative humidity is kept at 37% to prevent alterations in the bakelite resistivity [7]. The operating high voltages are optimized for each chamber and range from 10.0 to 10.4 kV [8]. The signal is picked up inductively on both sides of the detector by means of orthogonal copper strips (1, 2 and 4 cm wide), in the horizontal (bending plane)

[1] $y = \frac{1}{2} \ln \left(\frac{E + p_L}{E - p_L} \right)$, where E is the energy of a particle and p_L is the component of the momentum along the beam axis.

and vertical (non-bending plane) directions, one for each polarity. Strips are placed outside the electrodes. With the present FEE (ADULT ASIC [9]) the signal is discriminated without pre-amplification, with thresholds set at 7 mV and the mean charge per hit produced in the gas gap is of the order of 100 pC [10].

In those conditions, the instantaneous counting rate limit is 50 hits $\cdot s^{-1} \cdot cm^{-2}$, including some safety margins. Furthermore, the safe operation of the RPCs cannot be guaranteed for a cumulative charge larger than 50 mC/cm^2 (i.e. 500 Mhits/cm^2 in maxi-avalanche) [6]. These two limitations are not compatible with the expected conditions after the second LHC long shutdown (2019–2020). Indeed, peaks of counting rates up to 125 hits/s/cm^2 and a cumulative charge of 100 mC/cm^2 should be reached by the most exposed RPCs in central Pb–Pb collisions.

A possible solution for overcoming these two problems is to operate the RPCs at lower gain, like in ATLAS [11, 12] and in CMS [13, 14]. This requires a new electronics with amplification to reduce the charge per hit. The proposed solution is to replace the present front-end boards with new ones based on the FEERIC ASIC (Front-End Electronics Rapid Integrated Circuit) developed at the LPC in Clermont-Ferrand (France) [15]. Unlike ADULT, FEERIC performs amplification of the analog signals from the RPC before discrimination. This should allow to reduce by a factor 3–5 the charge produced in the gas, hence limiting the aging effects.

2. First results with cosmic rays

Preliminary measurements have been performed with around 10 FEERIC board prototypes for strips of 2 cm wide at the beginning of 2015, using the test bench available at INFN in Torino (Italy) [16]. The RPC under test has been filled with the same gas mixture and relative humidity adopted in the experiment.

Efficiency curves with FEERIC for various discrimination thresholds are compared in the left panel of figure 1, to the one obtained with ADULT with 7 mV threshold. The rise to the efficiency plateau is found to be quite sharp and a clear shift towards lower high voltage working points can be observed with FEERIC. The HV shift depends strongly on the FEERIC discrimination threshold, which is limited by the environmental noise conditions. The HV shift is found to be, however, never less than 600–700 V.

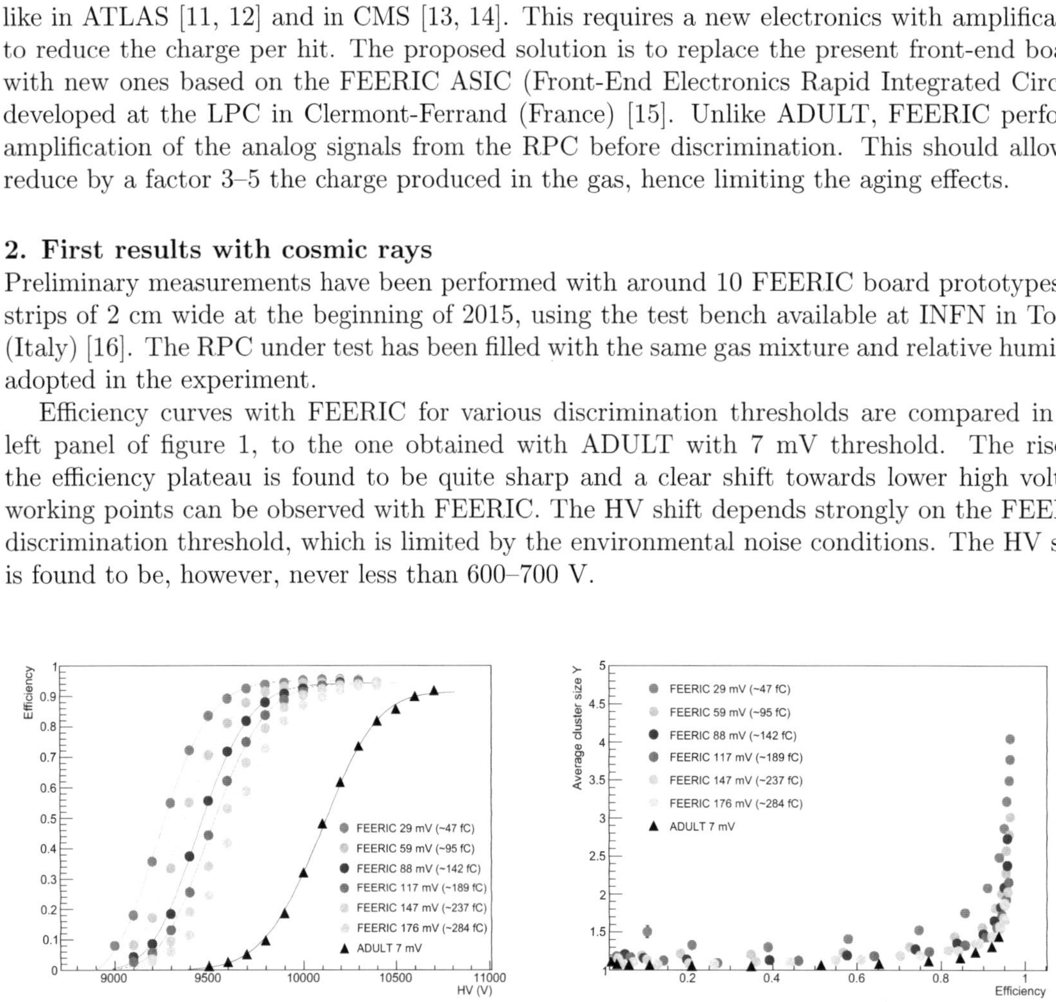

Figure 1. Efficiency curve (left) and cluster size measurements for strips of 2 cm wide (right) from cosmic data taking. Colors refer to different discrimination thresholds.

The cluster size as a function of the RPC efficiency is shown in figure 1 (right). In general the average value measured at the working point with FEERIC is slightly higher than what is measured with ADULT. This can be explained by considering the different signal amplitude distributions in the two working modes.

3. FEERIC in ALICE

From the results described above, it is however very difficult to anticipate the achievable charge reduction in the experimental working conditions. This is why it was foreseen to equip one of the 72 RPCs in the ALICE cavern with FEERIC electronics, in order to evaluate the performance during LHC Run 2. In February 2015, 39 cards were installed on one RPC with strips of 2 and 4 cm, placed in the fourth plane of the muon trigger. The performance in terms of efficiency, cluster size, current, single counting rate, stability and robustness are now evaluated from the data taking in pp collisions at $\sqrt{s} = 5$ and 13 TeV and in Pb–Pb collisions at $\sqrt{s_{\rm NN}} = 5$ TeV.

3.1. Working point determination

The operating point of the RPC equipped with FEERIC cards has been determined by means of a HV scan performed in June 2015 during low-luminosity pp runs at 13 TeV, for three thresholds. The results are shown in figure 2 (left). Full efficiency is reached for all threshold values and the results are very similar for positive (bending plane) and negative (non-bending plane) polarities.

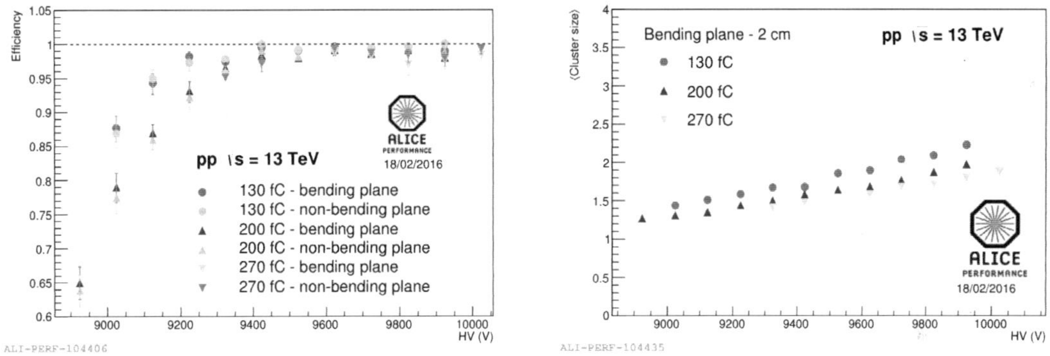

Figure 2. Efficiency plateau (left) and cluster size measurements for strips of 2 cm (right) wide in pp collisions at 13 TeV. Colors refer to different discrimination thresholds.

The HV dependence of the cluster size for strips of 2 cm is also shown in figure 2 (right). The average size of cluster is larger for the lower threshold values and increases with HV. As already noticed during the tests with cosmic rays, in maxi-avalanche mode (with ADULT cards with a threshold of 7 mV) the cluster size is smaller, ~1.4 strips for strips of 2 cm wide [10].

By considering these results and the fact that no noise is observed in the ALICE cavern for a threshold voltage above 33 mV (~60 fC) for all FEERIC card formats and polarities, the working point has been set at 9375 V for a 70 mV (~130 fC) threshold to make sure to be above the noise value. For comparison, the working point of this RPC when equipped with ADULT cards was 10125 V for a threshold of 7 mV (maxi-avalanche mode without FEE amplification). The HV difference is of 750 V, in line with what was observed during preliminary tests with cosmic rays.

3.2. Efficiency stability

Figure 3 (left) shows the time dependence of the efficiency of the RPC equipped with FEERIC. Measurements demonstrate a very high efficiency (~97%) for both bending and non-bending directions. The stability is also evident for the three collision systems characterizing the data taking.

The comparison with the other chambers is shown in figure 3 (right). It displays the efficiency of the other 71 RPCs still equipped with the non-amplified electronics and the chamber with

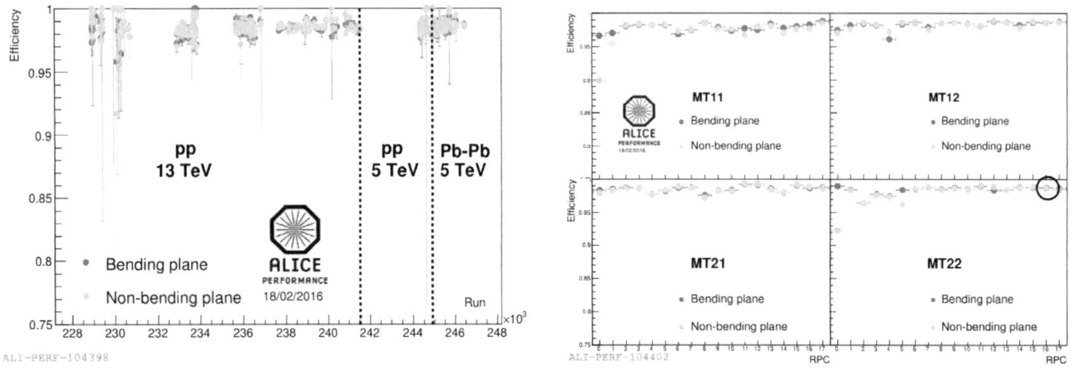

Figure 3. Left: efficiency of the RPC equipped with FEERIC measured in 2015. Right: average efficiency of the 72 RPCs composing the ALICE muon trigger. The circle indicates the chamber equipped with the new electronics. Red (green) points indicate the bending (non-bending) plane.

the new front-end in the circle. The values are averaged over the whole 2015. The efficiency of the chamber with FEERIC is perfectly in agreement with the results obtained for the other RPCs or slightly better.

3.3. Dark current and dark rate time dependence

Dark currents have been measured with dedicated runs in the absence of collisions, with the HV at their nominal values. The average dark current of the 71 RPCs equipped with ADULT at nominal voltage ranges from 2 μA to 4 μA, increasing with time as can be seen in figure 4 (left). Conversely, the dark current of RPC with FEERIC is much lower ($\sim$0.6 μA) and stable throughout the 2015 data taking, as shown in the right part.

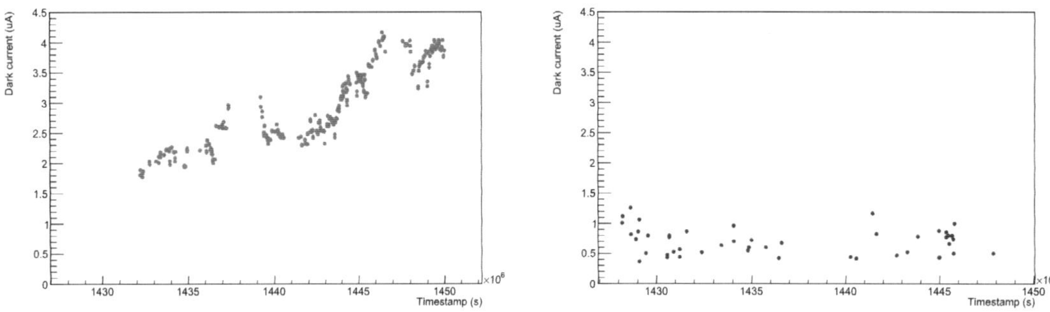

Figure 4. Average dark current drawn by the 71 RPCs equipped with ADULT (left) compared to the dark current of the RPC with FEERIC (right).

The dark rate has also been measured. This measurement is obtained from the strip scalers and hence it is dead time free. The average dark rate of RPCs with ADULT is below 0.05 Hz/cm^2 (figure 5, left) and stable in time, while the dark rate of RPC with FEERIC is larger, but usually below 0.1 Hz/cm^2.

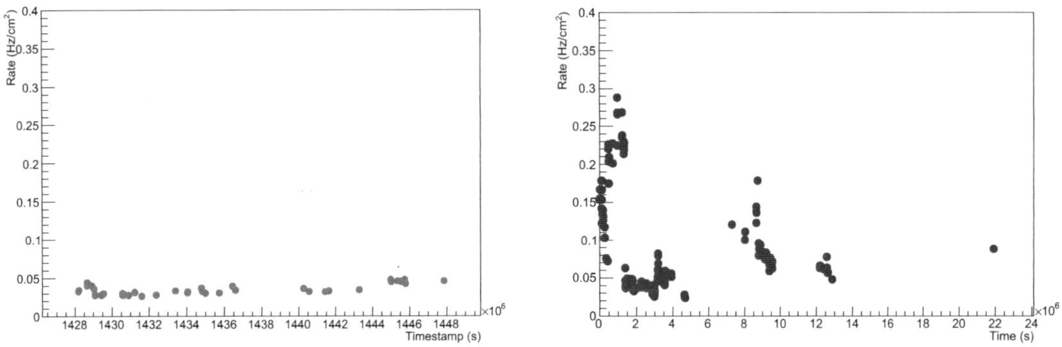

Figure 5. Average dark rate of the RPCs with ADULT (left) compared to the dark rate of the RPC with FEERIC (right). Both measurements refer to 2015 data taking.

3.4. Evaluation of the charge reduction with FEERIC

As previously mentioned, the main goal of this front-end electronics upgrade is to reduce the total charge produced in the gas gap, in order to prevent aging effects and to improve the rate capability. In order to evaluate the gain in charge obtained in the new working conditions, the RPC current has been measured at fixed interaction rate of ~400 kHz in July 2015 after subtraction of the dark component. Measurements have been performed at the HV value of 9375 V, corresponding to the working point with FEERIC, and at the HV value of 10125 V (ADULT working point). The results are shown in figure 6, together with two intermediate HV measurements. The gain in current (and therefore in charge) obtained is more than a factor 4.

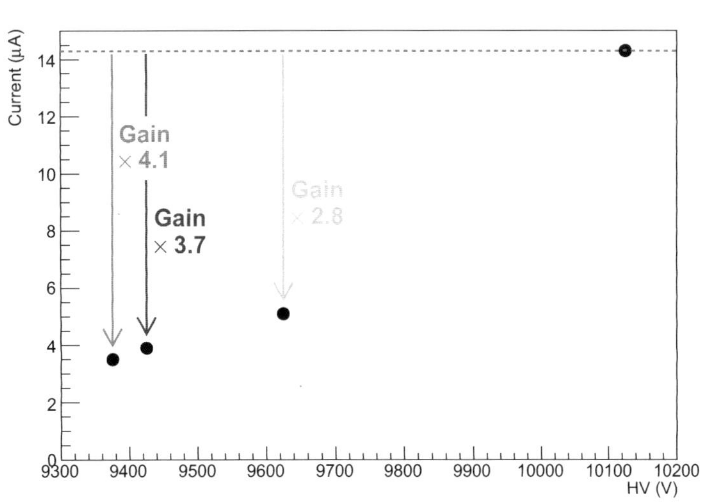

Figure 6. Current drawn in pp collisions by the RPC equipped with FEERIC at different HV. The gain with respect to the old working point is also merked by colored arrows.

4. Conclusions and perspectives

In summary, a completely new front-end electronics has been designed for the RPCs equipping the ALICE Muon Trigger system with the aim to increase the rate capability and limit aging effects. The performance of one RPC equipped with the new cards have been studied in a dedicated cosmic data taking and in pp and Pb–Pb collisions. It has been possible to lower the working point by 750 V maintaining the same efficiency larger than 95% measured with ADULT. The main goal of the FEERIC upgrade project has been achieved: a reduction in current (and therefore in total charge) of a factor 4 has been reached as compared to the present operations in maxi-avalanche. Performance related to cluster size and single counting rate are also within specifications. This study confirmed the robustness and the reliability on a long time scale of FEERIC in realistic conditions.

References

[1] ALICE Collaboration, The ALICE experiment at the CERN LHC, JINST 3 (2008) S08002.
[2] ALICE Collaboration, The forward muon spectrometer of ALICE: addendum to the technical proposal for a Large Ion Collider experiment at the CERN LHC, CERN–LHCC–96–032.
[3] R. Arnaldi et al., Study of the resistive plate chambers for the ALICE Dimuon Arm, Nucl. Instr. and Meth. A 456 (2000) 73.
[4] R. Santonico et al., Development of resistive plate counters, Nucl. Instr. and Meth. A 187 (1981) 377.
[5] R. Arnaldi et al., A low-resistivity RPC for the ALICE dimuon arm, Nucl. Instr. and Meth. A 451 (2000) 462.
[6] R. Arnaldi et al., Beam and aging tests with a highly-saturated avalanche gas mixture for the ALICE p-p data taking, Nucl. Phys. B 158 (2006) 149.
[7] R. Arnaldi et al., Aging tests and chemical analysis of Resistive Plate Chambers for the trigger of the ALICE dimuon arm, Nucl. Instr. and Meth. A 533 (2004) 112.
[8] M. Gagliardi, Commissioning and first performance of the resistive plate chambers for the ALICE muon arm, Nucl. Instr. and Meth. A 661 (2012) S45.
[9] R. Arnaldi et al., Front-end electronics for the RPCs of the ALICE dimuon, IEEE Trans. Nucl. Sci. 52 (2005) 1176.
[10] F. Bossù et al., Performance of the RPC-based ALICE muon trigger system at the LHC, JINST 7 (2012) T12002.
[11] ATLAS Collaboration, The ATLAS Experiment at the CERN Large Hadron Collider, JINST 3 (2008) S08003.
[12] F. Giannini et al., An 8 channel GaAs IC front-end discriminator for RPC detectors, Nucl. Instr. and Meth. A 432 (1999) 440.
[13] CMS Collaboration, The CMS experiment at the CERN LHC, JINST 3 (2008) S08004.
[14] M. Abbrescia et al., New developments on front-end electronics for the CMS Resistive Plate Chambers, Nucl. Instr. and Meth. A 456 (2000) 143.
[15] P. Dupieux et al., Upgrade of the ALICE muon trigger electronics, JINST 9 (2014) C09013.
[16] R. Arnaldi et al., Final results of the tests on the resistive plate chambers for the ALICE muon arm, Nucl. Instr. and Meth. A 602 (2009) 740.

IOP Conf. Series: Journal of Physics: Conf. Series **889** (2017) 012012 doi:10.1088/1742-6596/889/1/012012

Evolution of the gauge couplings and Weinberg angle in 5-dimensions for a G_2 gauge group

Mohammed Omer Khojali[a,1], **A. S. Cornell**[a,2], **Aldo Deandrea**[b,3] **and Giacomo Cacciapaglia**[b,4]

[a] National Institute for Theoretical Physics; School of Physics, University of the Witwatersrand, Wits 2050, South Africa

[b] Universite de Lyon, F-69622 Lyon, France; Universite Lyon 1, CNRS/IN2P3, UMR5822 IPNL, F-69622 Villeurbanne Cedex, France

E-mail: [1]khogali11@gmail.com, [2]alan.cornell@wits.ac.za, [3]deandrea@ipnl.in2p3.fr, [4] g.cacciapaglia@ipnl.in2p3.fr

Abstract. We explicitly test, in a simplified 5-dimensional model with G_2 gauge symmetry, the evolution of the gauge couplings. We assume that all the matter fields are propagating in the bulk, and consider orbifolds based on Abelian discrete groups which lead to 5-dimensional gauge theories compactified on an S^1/Z_2. The gauge couplings evolution is derived at one-loop level and used to test the impact on lower energy observables, in particular the Weinberg angle.

1. Introduction

Note that in coming years a new era of unexpected and exciting discoveries in high energy physics. For example, the CERN Large Hadron Collider (LHC) has been operating for some time and much experimental data has already been collected [1]. So far, the greatest achievement of the LHC has been the discovery of the missing building block of the SM, the Higgs particle (or, at least, a particle which most likely is the SM Higgs particle). On the other hand, no direct evidence of new physics beyond the SM has been found, yet. However, there are many reasons to believe that new physics should in fact show-up at, or about, the TeV scale [2, 3].

Theories which contain light elementary scalars can look unnatural because their masses receive quadratically divergent loop corrections, this problem is known as the hierarchy problem of the Standard Model (SM), as in this case the masses are pushed up to a cutoff scale [4]. In order to solve this problem we need to lower the cutoff scale (as was done in the case of large extra dimensional models), or to embed the Higgs field in a multiplet of a symmetry group larger than the 4-dimensional Poincare group (such as supersymmetry (SUSY)) [5]. As we know SUSY is a space-time symmetry mapping particles and fields of integer spin (bosons) into particles and fields of half integer spin (fermions), and vice-versa. Aside from SUSY there are other extensions of the 4-dimensional Poincare group, the most natural such choice would be to use the Poincare group of a higher dimensional gauge theory [6, 7]. A gauge theory defined in more than four dimensions can have many attractive features, one of these features is that the interactions at low energies may be truly unified and some of the distinct fields in four dimensions can be integrated in a single multiplet in higher dimensions, Higgs fields could be a part of the gauge fields. Another feature is the topology and structure of the extra-dimension may provide us with new ways of breaking symmetries, accounting for, at the same time, the hierarchy problem [8]. As such the SM Higgs field may originate from extra components of a higher dimensional gauge field. We therefore plan to discuss the evolution equations of the gauge couplings and Weinberg angle for a G_2 gauge group.

The structure of this paper is as follows: In section 2 we outline the G_2 model construction, in section 3 we explore the evolution of the gauge couplings and Weinberg angle for a G_2 gauge group. Our results and discussions will be given in section 4, In section 5 we conclude.

2. The G_2 Model Construction

In order to build a successful model we are first required to find a gauge group that contains $SU(2)_L\times U(1)_Y$ (and optionally the strong $SU(3)_c$), where the adjoint representation contains a doublet of SU(2) to be identified with the Higgs doublet. The second step is to normalise the U(1) gauge coupling so that the candidate Higgs has the correct hypercharge.

In this section we will explore a simplest five-dimensional gauge Higgs unification scenarion, where we use the gauge symmetry to be G_2. The extra dimension is compactified on a circle of radius R with a Z_2 orbifolding [9]. This orbifold is given as $Z_2 : y \to -y$, so our physical space is in the interval $y \in [0, \pi R]$ and has two fixed points at $y = 0$ and $y = \pi R$ [10, 11]. We assume that all matter fields are propagating in the bulk. The gauge bosons arise from the 4-dimensional components of the 5-dimensional gauge fields, whilst the Higgs field arises from the internal components of the gauge group G_2 compactified on an S^1/Z_2 orbifold [13, 14, 3]..

The G_2 gauge group contains $SU(3)$ as his maximal subgroup, and the decomposition under $SU(3)$ is given by:

$$\mathbf{14} = \mathbf{8} + \mathbf{3} + \bar{\mathbf{3}}, \tag{1}$$

form this we can see that there are two possible doublets, one contained in the adjoint of $SU(3)$, and the other in the triplets, where in the first case $g_1 = \sqrt{3}\, g_2$ while in the other case $g_1 = g_2/\sqrt{3}$. The decomposition under $SU(2) \times U(1)$ is given by:

$$14 = \left(3_0 + (2 + \bar{2})_{\sqrt{3}/2} + 1_0\right) + (2 + \bar{2})_{1/2\sqrt{3}} + (1 + \bar{1})_{-1/\sqrt{3}} \tag{2}$$

The other maximal subgroup is $SU(2) \times SU(2)$ under which:

$$\mathbf{14} = (\mathbf{1},\, \mathbf{3}) + (\mathbf{3},\, \mathbf{1}) + (\mathbf{2},\, \mathbf{4}) \tag{3}$$

IOP Conf. Series: Journal of Physics: Conf. Series **889** (2017) 012012 doi:10.1088/1742-6596/889/1/012012

where in this case the first $SU(2)$ has to be identified with the one contained in the $SU(3)$ gauge group, and then in this case, we can perform two possible way for breaking the group, either it is the rotation, which is now breaks $G_2 \to SU(3)$, and hence the glide to $SU(2) \times SU(2)$ or the viceversa. The fundamental representations under $SU(3)$ decompose as:

$$\mathbf{7} = \mathbf{3} + \bar{\mathbf{3}} + \mathbf{1} \tag{4}$$

and the fundamental representations under $SU(2) \times U(1)$ decompose as:

$$7 = (2 + \bar{2})_{1/2\sqrt{3}} + (1 + \bar{1})_{-1/\sqrt{3}} + 1_0 \tag{5}$$

We need the renormalization group equations (RGEs) to fill in the space between the predictions of the model at $\mu \gg M_Z$ and the experimental ones at $\mu \leq M_Z$. We can describe the contributions from the SM and Kaluza-Klein modes (KK) to the beta-functions in two separate terms, they are different and independent [12, 15].

3. The evolution of the gauge couplings and Weinberg angle for a G_2

The evolution of the gauge couplings in 4-dimensions for the G_2 gauge group at one-loop is given by:

$$16\pi^2 \frac{dg_i}{dt} = b_i^G \, g_i^3, \tag{6}$$

where the numerical coefficients in equation (6) are given by:

$$b_i^G = \left[\frac{53}{6}, -\frac{21}{6}, -\frac{63}{6}\right]. \tag{7}$$

We can then rewrite equation (6) in terms of α_i^{-1} as follows

$$\frac{1}{\alpha_i(\mu)} \frac{d \ln \alpha_i(\mu)}{d \ln \mu} = \frac{b_i}{2\pi}. \tag{8}$$

The one-loop beta functions for the gauge couplings in 5-dimensions for the G_2 gauge group are given by:

$$16\pi^2 g_3^{-3} \frac{d\, g_3}{d\, t} = -(S(t) - 1)\left(\frac{14}{6}\right), \tag{9}$$

$$16\pi^2 g_2^{-3} \frac{d\, g_2}{d\, t} = (S(t) - 1)\left(\frac{7}{24}\right), \tag{10}$$

$$16\pi^2 g_1^{-3} \frac{d\, g_1}{d\, t} = (S(t) - 1)\left(\frac{35}{8}\right), \tag{11}$$

where $S(t) = M_Z\, R e^t$ is the number of KK states, $t = \ln\left(\frac{\mu}{M_Z}\right)$ is the energy scale parameter, for $M_Z < \mu < 1/R$. We have chosen the Z boson mass as the renormalization point, that is when the energy $\mu = 1/R$ or $S(t) = 1$, in this case the whole beta-function reduces to the normal beta-functions [16, 17].

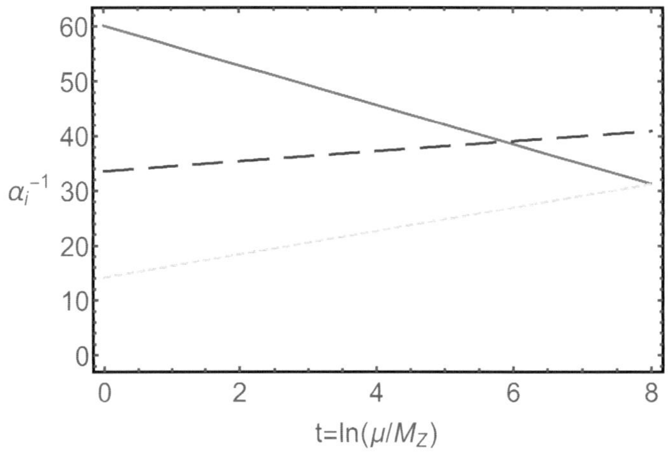

Figure 1. Running of α_i^{-1} in 4-dimensions for one-loop beta-functions, where α_3^{-1} is the (green) dotted line, α_2^{-1} is the (blue) dashed line and α_1^{-1} is the (red) solid line.

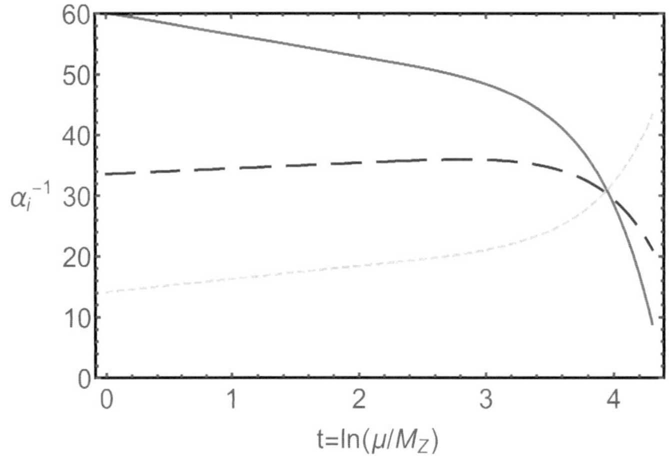

Figure 2. The evolution of theof α_i^{-1} in 5-dimensions for one-loop beta-functions, for the G_2 gauge group, where α_3^{-1} is the (green) dotted line, α_2^{-1} is the (blue) dashed line and α_1^{-1} is the (red) solid line.

4. Results and discussions

For our numerical calculation we choose the initial values for the gauge couplings based on the renormalisation point M_Z scale as: $g_1(M_Z) = 0.46597$, $g_2(M_Z) = 0.63337$ and $g_3(M_Z) = 1.02739$. In Figure 1 , we present the evolution of α_i^{-1} in 4-dimensions for the one-loop beta-functions, in this figure we see that α_1^{-1} and α_3^{-1} are approximately unified at some energy scale $t \sim 8.0$. Whilst Figure 2 , shows the evolution of the α_i^{-1} in 5-dimension for the one-loop beta-function for the G_2 gauge group. From this one can see that $\alpha_1^{-1}, \alpha_2^{-1}$ and α_3^{-1} are approximately unified at $t \sim 4.0$.

In Figure 3 , we present the evolution of the Weinberg angle for the one-loop beta-functions, for different values of compactification scale, as example, for $R^{-1} = 10\,\text{TeV}$,

$\sin^2\theta_W \sim 0.42$ at $t \sim 5.56$. When the fifth dimension KK modes become kinematically accessible, there are large changes in the Weinberg angle up until we reach the cut-off scale. We have chosen the cut-off for our effective theory, as $g_1 = g_2$, as shown in Table 1.

Scenario	$t(R_1)$	$t(R_2)$	$t(R_3)$
5D G_2	4.529	5.045	5.245

Table 1. The cut-offs in 5 dimensions for the G_2 gauge group for three different compactification radii $R^{-1} = 1, 5$ and 10 TeV, where $t = \ln(\mu/M_Z)$.

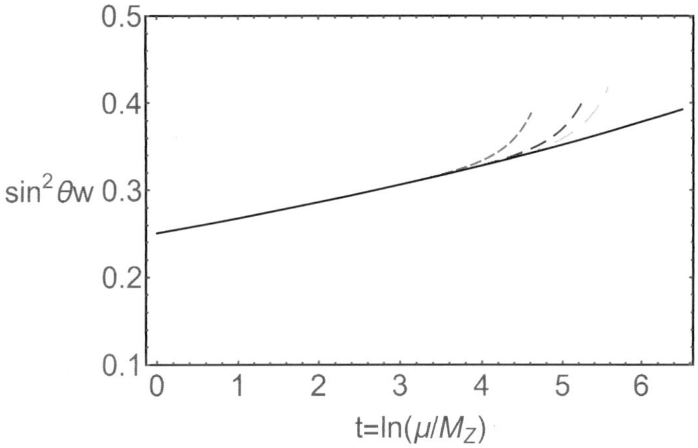

Figure 3. The evolution of the Weinberg angle $\sin^2\theta_W$ for the one-loop beta-function, for different values of compactification scales, $R^{-1} = 1$ TeV (red), $R^{-1} = 5$ TeV (blue) and $R^{-1} = 10$ TeV (green) as a function of t.

5. Conclusion

In conclusion, in this paper we explicitly tested, in a simplified 5-dimensional model with a G_2 gauge symmetry, the evolution of the gauge couplings and Weinberg angle. We observed that when the fifth dimension KK-modes switch on all the previous physical observables evolution change rapidly.

Acknowledgments

This work is based on the research supported by the National Research Foundation of South Africa.

References

[1] G. Cacciapaglia and B. Kubik, JHEP **1302**, 052 (2013) [arXiv:1209.6556 [hep-ph]].

IOP Conf. Series: Journal of Physics: Conf. Series **889** (2017) 012012 doi:10.1088/1742-6596/889/1/012012

[2] A. Abdalgabar, A. S. Cornell, A. Deandrea and A. Tarhini, Phys. Rev. D **88**, 056006 (2013) [arXiv:1306.4852 [hep-ph]].
[3] G. Bhattacharyya, A. Datta, S. K. Majee and A. Raychaudhuri, Nucl. Phys. B **760**, 117 (2007) [hep-ph/0608208].
[4] C. Csaki, C. Grojean and H. Murayama, Phys. Rev. D **67**, 085012 (2003) [hep-ph/0210133].
[5] N. S. Manton, Nucl. Phys. B **158**, 141 (1979).
[6] G. R. Dvali, S. Randjbar-Daemi and R. Tabbash, Phys. Rev. D **65**, 064021 (2002) [hep-ph/0102307].
[7] I. Antoniadis and K. Benakli, Phys. Lett. B **326**, 69 (1994) [hep-th/9310151].
[8] N. Haba, M. Harada, Y. Hosotani and Y. Kawamura, Nucl. Phys. B **657**, 169 (2003) Erratum: [Nucl. Phys. B **669**, 381 (2003)] [hep-ph/0212035].
[9] I. Antoniadis, K. Benakli and M. Quiros, Phys. Lett. B **331**, 313 (1994) [hep-ph/9403290].
[10] L. J. Hall, H. Murayama and Y. Nomura, Nucl. Phys. B **645**, 85 (2002) [hep-th/0107245].
[11] A. Deandrea, J. Welzel, P. Hosteins and M. Oertel, Phys. Rev. D **75**, 113005 (2007) [hep-ph/0611172].
[12] T. Ohlsson and S. Riad, Phys. Lett. B **718**, 1002 (2013) [arXiv:1208.6297 [hep-ph]].
[13] I. Antoniadis, K. Benakli and M. Quiros, Phys. Lett. B **331**, 313 (1994) [hep-ph/9403290].
[14] N. Kitazawa and Y. Sakai, Mod. Phys. Lett. A **31**, no. 07, 1650041 (2016) [arXiv:1509.03409 [hep-ph]].
[15] K. R. Dienes, E. Dudas and T. Gherghetta, Phys. Lett. B **436**, 55 (1998) [hep-ph/9803466].
[16] A. S. Cornell, A. Deandrea, L. X. Liu and A. Tarhini, Mod. Phys. Lett. A **28**, no. 11, 1330007 (2013) [arXiv:1209.6239 [hep-ph]].
[17] L. X. Liu and A. S. Cornell, Phys. Rev. D **86**, 056002 (2012) [arXiv:1204.0532 [hep-ph]].

IOP Conf. Series: Journal of Physics: Conf. Series **889** (2017) 012013 doi:10.1088/1742-6596/889/1/012013

Studies of the b-tagged control region for same sign $W^{\pm}W^{\pm}$ production in proton-proton collision data recorded by the ATLAS detector

Raynette van Tonder

Department of Physics, University of Cape Town, Private Bag X3, Rondebosch 7701, South Africa

E-mail: VTNRAY001@myuct.ac.za

Abstract. Vector Boson Scattering has been identified as a promising process in order to study the the nature of electroweak symmetry breaking. One of its production mechanisms, same electric charge W boson scattering is a rare Standard Model process. At the LHC, W boson scattering can occur when W bosons are radiated off incoming proton beams and scatter. The proton's remnants would be detected in the forward calorimeters of the detector as jets, while events are selected where both W bosons decay leptonically. This process therefore has a distinct experimental signature of two same electric charge leptons and two jets, however various Standard Model processes can mimic this signature. These background processes are modelled by making use of Monte Carlo simulations and then tested in several same electric charge dilepton control regions. Non-prompt leptons originating from $t\bar{t} \rightarrow W^{+}bW^{-}b \rightarrow l^{+}l^{-}\nu\nu jjb\bar{b}$ are tested in the b-tagged control region, which requires that at least one of the two jets is identified as a b jet. In this presentation results from the b-tagged control region, using proton-proton collision data at $\sqrt{s} = 13$ TeV recorded by the ATLAS detector in 2015, will be discussed.

1. Introduction

An interesting feature arises with the calculation of the scattering amplitude for WW scattering. In the absence of a Standard Model (SM) Higgs boson, the longitudinally polarised scattering amplitude grows as a function of center-of-mass energy squared and violates unitarity at $\sqrt{s} \approx 1$ TeV [1]. With the addition of a Higgs scalar the scattering amplitude of these processes can be regulated at high energies, restoring unitarity, only if the recently discovered Higgs boson [2] [3] behaves as the predicted SM Higgs boson [4]. Representative Feynman diagrams of the processes are shown in Figures 1 and 2. One type of Vector Boson Scattering, W boson scattering, is therefore a process of great interest, since the process is linked to the mechanism of electroweak symmetry breaking. Two possible processes may be used in order to study WW scattering: same electric charge WW scattering, $W^{\pm}W^{\pm}$, and opposite electric charge WW scattering, $W^{\mp}W^{\pm}$. Even though the predicted electroweak cross section for the opposite sign $W^{\mp}W^{\pm}$ process is far greater than the electroweak cross section for the same sign $W^{\pm}W^{\pm}$ process, the strong scattering cross section completely dominates the former process. On the other hand, the QCD processes (mediated by a gluon) for the $W^{\pm}W^{\pm}$ process are very small. This is due to the fact that LO gluon-gluon initial states are not present in the $W^{\pm}W^{\pm}$ process [5], which makes the electroweak and strong cross sections for this process roughly of the same order [6]. The

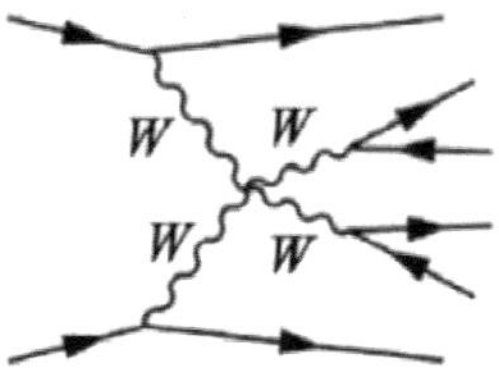

Figure 1. Feymnan diagram of WW scattering, which violates unitarity at $\sqrt{s} \approx 1\,\text{TeV}$.

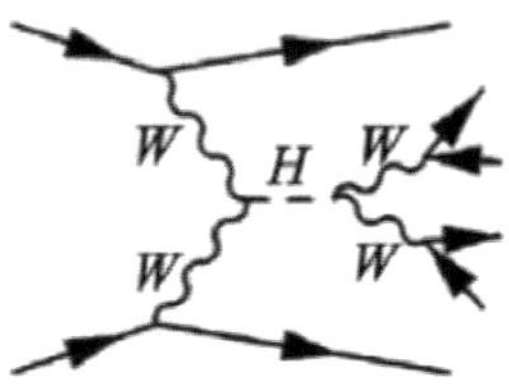

Figure 2. The addition of a Higgs scalar to the Feynman diagram of WW scattering restores unitarity.

excellent signal-to-background ratio together with the unique experimental signature of same sign $W^{\pm}W^{\pm}$ scattering, which greatly reduces the Standard Model background processes, makes this process ideal for studying Vector Boson Scattering and probing the nature of electroweak symmetry breaking.

2. Same sign $W^{\pm}$ production

An incredibly rare process, same electric charge W boson scattering can occur at hadron colliders like the LHC, as an interaction of W bosons that are radiated off incoming proton beams. These W bosons scatter and subsequently decay, however events selected are where the W bosons decay leptonically, i.e $W^{\pm} \rightarrow l^{\pm}\nu$, $l = e, \mu$. The process therefore has a very distinctive experimental signature of a lepton pair with the same electric charge and two high energy forward jets. Two analysis regions for the study of $W^{\pm}W^{\pm}jj$ production can be defined: an inclusive analysis region and a VBS analysis region [7]. The inclusive analysis region includes a combination of both strong and weak production mechanisms, which is used to measure the fiducial cross section of $W^{\pm}W^{\pm}$ production. In the VBS analysis region, events from strong $W^{\pm}W^{\pm}jj$ production are considered as background in order to study the purely electroweak $W^{\pm}W^{\pm}jj$ production process. The Standard Model cross section predictions for the inclusive analysis region and VBS analysis region are $1.52 \pm 0.11\,\text{fb}$ and $0.95 \pm 0.06\,\text{fb}$, respectively [7].

This rare Standard Model process has recently been observed by ATLAS and CMS, by making use of pp collision data at $\sqrt{s} = 8\,\text{TeV}$ collected by both detectors. Cross sections for both the inclusive analysis and VBS analysis regions were measured by ATLAS. The cross section for strong and electroweak $W^{\pm}W^{\pm}jj$ production in the inclusive analysis region was measured to be $\sigma^{fid} = 2.1 \pm 0.5\,\text{fb}$, while the cross section for electroweak $W^{\pm}W^{\pm}jj$ production, which includes interference of strong production, was measured to be $\sigma = 1.3 \pm 0.4\,\text{fb}$ for the VBS analysis region [7]. Furthermore, the observed combined significance over the background-only hypothesis was 4.5 and 3.6 standard deviations in the inclusive region and the VBS region, respectively. CMS measured the cross section for electroweak $W^{\pm}W^{\pm}jj$ production to be $\sigma = 4.0^{+2.4}_{-2.0}\,\text{fb}$ at an observed significance level of 2 standard deviations [8]. Results from both ATLAS and CMS can be seen in Figures 3 and 4, respectively, showing an excess of data events over the predicted background processes.

3. The b-tagged control region

Many SM processes can mimic the signal region for $W^{\pm}W^{\pm}jj$ production and must be excluded or reduced as background processes by defining specific event selections. In order to effectively reduce contributions from background processes, Monte Carlo (MC) simulations are used to estimate these background processes. The MC predictions must then be tested in several same charge di-lepton control regions. Control regions are regions used to test MC predictions of

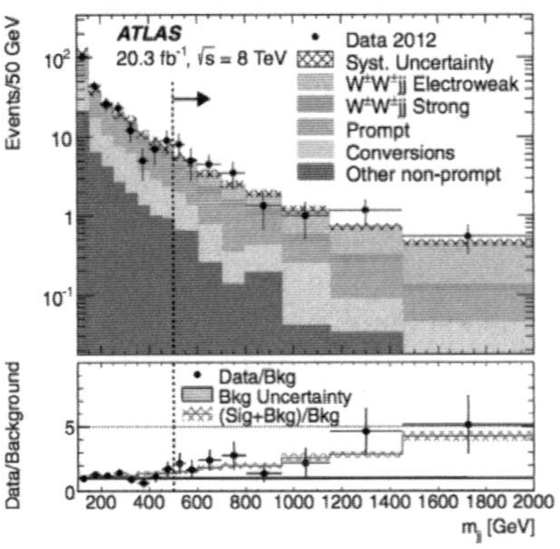

Figure 3. Dijet invariant mass distribution for the predicted background and signal events obtained by ATLAS. The hashed bars represent the systematic uncertainties in the total prediction, while the lower plot shows the ratio of the total prediction to background and the uncertainty [7].

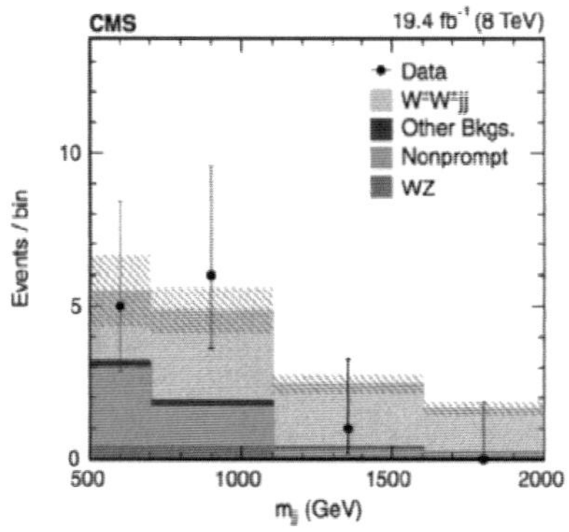

Figure 4. Dijet invariant mass distribution for predicted background and signal events obtained by CMS. The hashed bars represent the systematic and statistical uncertainties [8].

background processes by specifically selecting a region where a certain background process of interest is dominant. Non-prompt leptons originating from $t\bar{t} \to l\nu jjb\bar{b}$ are tested in the b-tagged control region, which requires that at least one of the two jets is identified as a b-jet [7]. By requiring a b-tagged jet in the event selection for this control region, events from non-prompt backgrounds originating from $t\bar{t}$ processes are specifically selected. This control region is also used to test the description of event kinematics for the charge misidentification background as well as the fake background, where jets are mis-reconstructed as charged leptons. Figure 5 shows the p_T distribution for the leading and sub-leading leptons in the b-tagged control region, from proton-proton collision data at $\sqrt{s} = 8\,\text{TeV}$ recorded by the ATLAS detector in 2012. In these plots it can be seen that this region is clearly dominated by fakes from $t\bar{t}$ and other non-prompt leptons that originated from hadronic decays. However $t\bar{t} + W/Z$ processes as well as the charge misidentification background also play important roles in this control region. It is important to understand the backgrounds in this control region in order to ensure that the lepton p_T defined in the event selection excludes contributions from non-prompt leptons originating from hadronic decays.

4. Conclusion and outlook

Vector Boson Scattering have been identified as a promising process in order to study and understand the mechanism of electroweak symmetry breaking. One of the production mechanisms of VBS is a rare Standard Model process known as same sign W boson scattering. First evidence of this process have recently been observed by both ATLAS and CMS, where both analyses made use of pp collision data at $\sqrt{s} = 8\,\text{TeV}$. Further studies implementing pp collision data collected by the ATLAS and CMS detectors during 2015 and 2016, at a higher center-of-mass energy than the previous analyses, will aim to provide conclusive evidence for the existence of electroweak $W^{\pm}W^{\pm}jj$ production, as well as increase our current understanding of VBS processes.

IOP Conf. Series: Journal of Physics: Conf. Series **889** (2017) 012013 doi:10.1088/1742-6596/889/1/012013

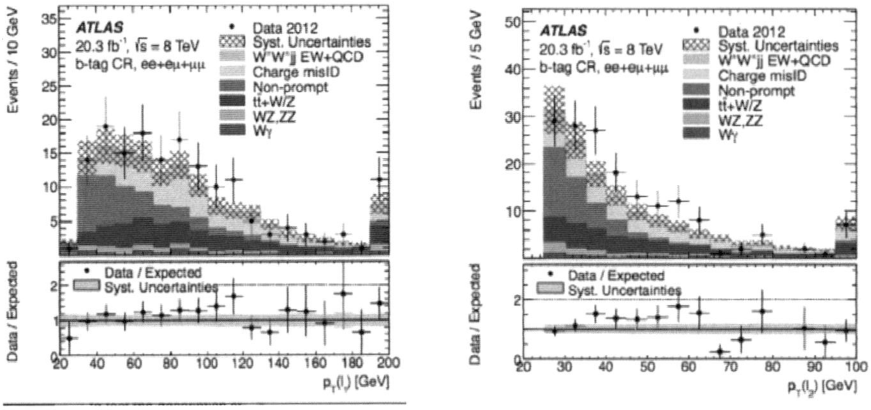

Figure 5. Transverse momentum distributions for leading (left) and sub-leading (right) leptons in the b-tagged control region. The black, hatched bars in the upper plots shows the systematic uncertainty on the total prediction, while the yellow band in the lower plots shows the uncertainty as a fraction of the total prediction [9].

Acknowledgments
I wish to express my sincere gratitude for the guidance of my supervisors, Dr. Sahal Yacoob and Dr. Andrew Hamilton. In addition, I would also like to thank the UCT Postgraduate Funding Centre for funding this research.

References
[1] Rindani S D 2009 *Physics at the Large Hadron Collider* (Springer) p 45
[2] ATLAS Collaboration 2012 *Phys. Lett. B* **716** 1–29
[3] CMS Collaboration 2012 *Phys. Lett. B* **716** 30–61
[4] Alboteanu A, Kilian W and Reuter J 2008 *J. High Energy Phys.* 010
[5] Accomando E, Ballestrero A *et al.* 2006 *J. High Energy Phys.* 093
[6] Zhu B, Govoni P *et al.* 2011 *Eur. Phys. J. C* **71** 1
[7] ATLAS Collaboration 2014 *Phys. Rev. Lett.* **113** 141803
[8] CMS Collaboration 2015 *Phys. Rev. Lett.* **114** 051801
[9] ATLAS Collaboration 2016 *arXiv preprint arXiv:1611.02428*

IOP Conf. Series: Journal of Physics: Conf. Series **889** (2017) 012014 doi:10.1088/1742-6596/889/1/012014

The anisotropy effect on jet using AdS/CFT Correspondence

R Morad[1,a] and K Bitaghsir Fadafan[2,b]

[a]Department of Physics, University of Cape Town, Private Bag X3, Rondebosch 7701, South Africa

[b]Department of Physics, Shahrood University of Technology, P.O.Box 3619995161 Shahrood, IRAN

E-mail: [1]razieh.morad@uct.ac.za, [2] bitaghsir@shahroodut.ac.ir

Abstract.

We study the collective flow effects on the jet of light quark in a strongly coupled plasma. We consider the finite temperature static strongly coupled anisotropic plasma which has been generated by a space dependent axion term. By computing the falling string solution in this background, we find that the jet quenching increases in both longitudinal and transverse direction in comparison with the static isotropic medium. Although, the enhancement of quenching is larger in the beam direction.

1. Introduction

The experiments at the Relativistic Heavy Ion Collider (RHIC) and the Large Hadron Collider (LHC) have produced a strongly-coupled matter which is a deconfined state of QCD, Quark-Gluon Plasma (QGP) [1, 2, 3, 4]. While lattice QCD is the proper tool for understanding the static equilibrium thermodynamics of such strongly coupled plasma, it does not allow us to calculate its dynamical evolution in heavy-ion collision. Recently, a novel tool called "the AdS/CFT correspondence" [5, 6, 7, 8, 9] provides valuable insight into the strongly coupled plasma.

Although ideal hydrodynamics successfully describes many properties of QGP, further studies propose that the medium must be anisotropic in the early times during the formation of QGP [10, 11, 12, 13, 14, 15]. These models proposed that within this time period, the pressure of the medium along the transverse direction may surpass the pressure in the longitudinal direction.

The anisotropic geometries have been studied using the gauge/gravity duality [16, 17]. In [16], the dual theory is an anisotropic $\mathcal{N} = 4$ SYM plasma with rotational symmetry in the $x - y$ directions, while the z direction corresponds to the beam direction. The drag force and jet quenching parameter in this anisotropic medium were calculated [18, 19, 20]. The results show that the energy loss of partons depends on the relative orientation between the anisotropic direction and the velocity of the quark. Also, pQCD calculations on jet energy loss show that the resulting medium-induced gluon radiation does not depend solely on the energy density of the medium, but also on the collective flow. Their results show that flow effects lead to a characteristic breaking of the rotational symmetry of the average jet energy [21, 22]. In this paper, we studied the effect of anisotropy on light quark using the AdS/CFT correspondence. We found that the maximum stopping distance is very sensitive to the anisotropy parameter and is smaller in the direction of beam.

2. Anisotropic background

In this section we review the anisotropic background introduced in [16] which is the gravity dual to a deformation of $\mathcal{N}=4$ SYM by adding a θ term to the action that depends linearly on one of the three spatial coordinates z as $\theta = 2\pi\, n_{D7}\, z$, where n_{D7} can be thought as the density of D7-branes homogeneously distributed along the anisotropic direction. The type IIB supergravity solution presented in this paper is static and anisotropic with finite temperature. Moreover it is regular on and outside the horizon and asymptotically approaches AdS_5.

The supergravity solution in the string frame is given by [16]

$$
\begin{aligned}
ds^2 &= \frac{L^2}{u^2}\left(-\mathcal{F}\mathcal{B}dt^2 + dx^2 + dy^2 + \mathcal{H}dz^2 + \frac{du^2}{\mathcal{F}}\right) + L^2 e^{\phi/2} ds^2_{S^5} \\
\phi &= \phi(u), \qquad \chi = a\, z,
\end{aligned} \tag{1}
$$

where ϕ and χ, are the dilaton and axion fields respectively, and the anisotropy parameter is $a = \frac{g^2_{YM} n_{D7}}{4\pi}$. The metric functions are $\mathcal{F}, \mathcal{B}$ and $\mathcal{H}$ which depend on the holographic radial coordinate, u the radius of horizon, u_H and the anisotropy parameter, a. The boundary of space is at $u = 0$ and the horizon located at $u = u_H$ where the blackening function is zero, $\mathcal{F}(u_H) = 0$.

The metric functions are parameterized by two parameters, the dilaton field value at the horizon ϕ_h and u_H or equivalently by T and a/T. At small a/T, the entropy density scales as in the isotropic case $S_{iso} = \frac{\pi^2}{2} N_c^2 T^3$, while at large a/T it scales as $S_{aniso} = c_{ent} N_c^2 a^{1/3} T^{8/3}$, where c_{ent} is a constant. For small a/T the metric functions and the radius of the horizon are known as some expansions around the black D3-brane solution [16],

$$
\begin{aligned}
\mathcal{F} &= 1 - \frac{u^4}{u_H^4} + a^2\mathcal{F}_2 + \mathcal{O}(a^4), \\
\mathcal{B} &= 1 + a^2\mathcal{B}_2 + \mathcal{O}(a^4), \\
\log\mathcal{H} &= \frac{a^2 u_H^2}{4}\log\left(1 + \frac{u^2}{u_H^2}\right) + \mathcal{O}(a^4), \\
u_H &= \frac{1}{\pi T} + \frac{5\log 2 - 2}{48\pi^3 T^3}a^2 + \mathcal{O}(a^4),
\end{aligned} \tag{2}
$$

in which

$$
\mathcal{F}_2 = \frac{1}{24 u_H^2}\left[8u^2(u_H^2 - u^2) - 10u^4\log 2 + 3u_H^4 + 7\log\left(1 + \frac{u^2}{u_H^2}\right)\right], \tag{3}
$$

$$
\mathcal{B}_2 = -\frac{u_H^2}{24}\left[\frac{10u^2}{u_H^2 + u^2} + \log\left(1 + \frac{u^2}{u_H^2}\right)\right]. \tag{4}
$$

The pressures and energies which are useful on making a naive connection with the weak coupling results can be found from the expectation values of the stress energy tensor and read

$$
\begin{aligned}
&E = 3e_1 + a^2 e_2 + O(a^4)\,, \quad P_\perp = e_1 + a^2 e_2 + O(a^4)\,, \quad P_\parallel = e_1 - a^2 e_2 + O(a^4)\,, \\
&\text{where} \quad e_1 = \frac{N_c^2 T^2}{32} + O(a^4)\,, \quad e_2 = \frac{N_c^2 T^2}{32}
\end{aligned} \tag{5}
$$

where N_c is the number of colors. Notice that

$$
P_\parallel < P_\perp \tag{6}
$$

3. Holographic jets in anisotropic background

The goal of this section is to study the effect of anisotropy of the medium on a light quark. The physical setup of interest is one of a back-to-back jet pair created in a quark-gluon plasma. We therefore consider configurations for which the string is created at a point and expands in space-time such that the two endpoints of the string move away from each other; the total spatial momentum of the string vanishes. With an appropriate choice of coordinates, in the rest frame of the plasma (equivalent to the rest frame for the whole string) one half of the string has a large spatial momentum in the $+r$ direction in the $(x-z)$ plane with angle ϕ with respect to the z coordinate, while the other half of the string has a large spatial momentum in the $-r$ direction; in this case the embedding function of string $X^\mu(\tau,\sigma)$ will be a map to $(t(\tau,\sigma), x(\tau,\sigma), 0, z(\tau,\sigma), u(\tau,\sigma))$.

The profile of an open string that is created at a point in space at time $t=t_c$ is given by

$$t(0,\sigma)=t_c\,,\;\; x(0,\sigma)=0\,,\;\; z(0,\sigma)=0\,,\;\; u(0,\sigma)=u_c, \tag{7}$$

where $\sigma\in[0,\,\pi]$. After the creation at time t_c, the string evolves from a point into an extended object and the string endpoints fall toward the horizon. In the case of anisotropic background the extension of the string in the x direction differs from its extension in the z direction; see figure 1 for a visualization of the string profile at various times after creation in both x and z direction. For precise numerical studies of the string profile, it is more convenient to use the Polyakov action instead of the Nambu-Goto action [23, 24, 25]. With the Polyakov action, one introduces additional degrees of freedom into the problem by allowing a nontrivial worldsheet metric η^{ab}; with these additional degrees of freedom, one can make the equations of motion well-behaved everywhere on the worldsheet [23, 24, 25]. The Polyakov action for the string has the form

$$S_P=-\frac{T_0}{2}\int d^2\sigma\,\sqrt{-\eta}\,\eta^{ab}\,\partial_a X^\mu\partial_b X^\nu\,G_{\mu\nu}\,. \tag{8}$$

Variation of the Polyakov action with respect to the embedding functions X^μ leads to the equation of motion

$$\begin{aligned}\partial_a[\sqrt{-\eta}\,\eta^{ab}\,G_{\mu\nu}\,\partial_b X^\nu] &= \tfrac{1}{2}\sqrt{-\eta}\,\eta^{ab}\tfrac{\partial G_{\nu\rho}}{\partial X^\mu}\,\partial_a X^\nu\partial_b X^\rho\\ \Longleftrightarrow\qquad \nabla_a\,\Pi^a_\mu &= -\tfrac{T_0}{2}\,\eta^{ab}\tfrac{\partial G_{\nu\rho}}{\partial X^\mu}\,\partial_a X^\nu\partial_b X^\rho,\end{aligned} \tag{9}$$

where Π^a_μ are the canonical momentum densities associated with the string that are obtained from varying the action with respect to the derivatives of the embedding functions. In order to optimize the performance of the numerical integrator, we choose a worldsheet metric of the form [23, 24, 25]

$$\|\eta_{ab}\|=\begin{pmatrix}-\Sigma(x,u) & 0\\ 0 & \Sigma(x,u)^{-1}\end{pmatrix}, \tag{10}$$

where Σ is called a stretching function, which can be a function of $x(\tau,\sigma)$ and $u(\tau,\sigma)$. In fact, the choice of worldsheet metric is a choice of gauge. A common choice is conformal gauge with $\Sigma=1$. We choose Σ such that the singularities in the equations of motion are cancelled.

In order to solve the equations of motion we need a self-consistent initial conditions (IC) for the string profile which obey the constraint equation,

$$-\mathcal{F}(u)^2\mathcal{B}\,\dot{t}^2+\mathcal{F}(u)\,\dot{x}^2+\mathcal{H}\mathcal{F}\,\dot{z}^2+\dot{u}^2=0\,. \tag{11}$$

The σ derivatives of X^μ are initially zero for the string with point-like IC. So, in order to satisfy (11) we just need to choose IC that satisfy $\dot{X}^2=0$ and obey the boundary condition for open string. We seek IC such that the string is long-lived, has most of its energy and momentum concentrated near its endpoints, and produces stable numerical solutions. We choose the following IC for the string in this background

$$\begin{aligned}
\dot{x}(0,\sigma) &= A\,u_c\cos(\sigma)\sin(\phi)\,,\\
\dot{z}(0,\sigma) &= A\,u_c\cos(\sigma)\cos(\phi)\,,\\
\dot{u}(0,\sigma) &= u_c\sqrt{f(u_c)}\,(1-\cos 2\sigma)\,,\\
\dot{t}(0,\sigma) &= u_c\sqrt{\frac{A^2\cos^2(\sigma)\left(H(u_c)\cos^2(\phi)+\sin^2(\phi)\right)+(1-\cos(2\sigma))^2}{B(u_c)}}\,,
\end{aligned} \tag{12}$$

where u_c and A are free parameters that can be related to the energy and momentum of the dual quark in the field theory (see below). These IC yield a string profile that is symmetric about $r = 0$ at all times, because $\dot{x}(0,\sigma)$ and $\dot{z}(0,\sigma)$ are antisymmetric about $\sigma = \pi/2$ while $\dot{u}(0,\sigma)$ is symmetric.

We realize that the equations of motion remain well behaved everywhere on the worldsheet by choosing any stretching function of the following form,

$$\Sigma(x,u) = \left(\frac{1-u/u_H}{1-u_c/u_H}\right)^a\left(\frac{u_c}{u}\right)^b \tag{13}$$

Now we can solve the equation of motion Eq. (9) numerically with *Mathematica*'s `NDSolve` to obtain the embedding functions X^μ as a function of (τ,σ). We choose the values of a and b case by case; a and b are in the range of 1 to 3. The shape of a representative string solution for the value of $\phi = \pi/3$ at different times is depicted in figure 1. As expected, the two endpoints of the string move away from each other as the string extends along the $(x,\,z)$ direction and falls toward the horizon. Numerical results show that the string in the anisotropic direction is suppressed with respect to the isotropic direction even at small anisotropy parameter.

It is important to make a minor comment on the energy and momentum of the string. Since $G^{\mu\nu}$ depends only on u, for μ corresponding to $(t,\,\vec{x})$ we have

$$\nabla_a\Pi^a_\mu = 0. \tag{14}$$

Hence the corresponding momentum densities Π^a_μ are conserved Noether currents on the worldsheet associated with the invariance of the action under spacetime translations. The Π^a_μ describe the flow of the μ component of the spacetime momentum of the string along the a direction on the worldsheet [26]. So the total energy of the string at any time is constant and equal to the initial energy of the string

$$\begin{aligned}
E_{\rm string} &= -\int_0^\pi d\sigma\,\sqrt{-\eta}\,\Pi^\tau_t(0,\sigma)\\
&= \frac{\lambda}{2\pi}\int_0^\pi d\sigma\,\frac{\mathcal{F}(u_c)}{u_c\,\Sigma(u_c)}\sqrt{\frac{A^2\cos^2(\sigma)\left(H(u_c)\cos^2(\phi)+\sin^2(\phi)\right)+(1-\cos(2\sigma))^2}{B(u_c)}}\,.
\end{aligned} \tag{15}$$

Also, the total momentum in x and z direction is conserved and can be calculated by the string IC as follows

$$P_{q\,x} = \frac{\sqrt{\lambda}\,A\,\sin(\phi)}{2\pi\;u_c\,\Sigma(u_c)}\quad,\quad P_{q\,z} = \frac{\sqrt{\lambda}\,A\,\mathcal{H}(u_c)\,\cos(\phi)}{2\pi\qquad u_c\,\Sigma(u_c)}. \tag{16}$$

In the above integral, we used the fact that the total momentum of the string in $x - z$ direction is zero and it is completely symmetric about the point $\sigma = \pi/2$. So, the momentum of the quark is equal to the momentum of the anti-quark in the opposite direction. We can see that the two parameters A and u_c in the IC of the string determine the energy and momentum of the string

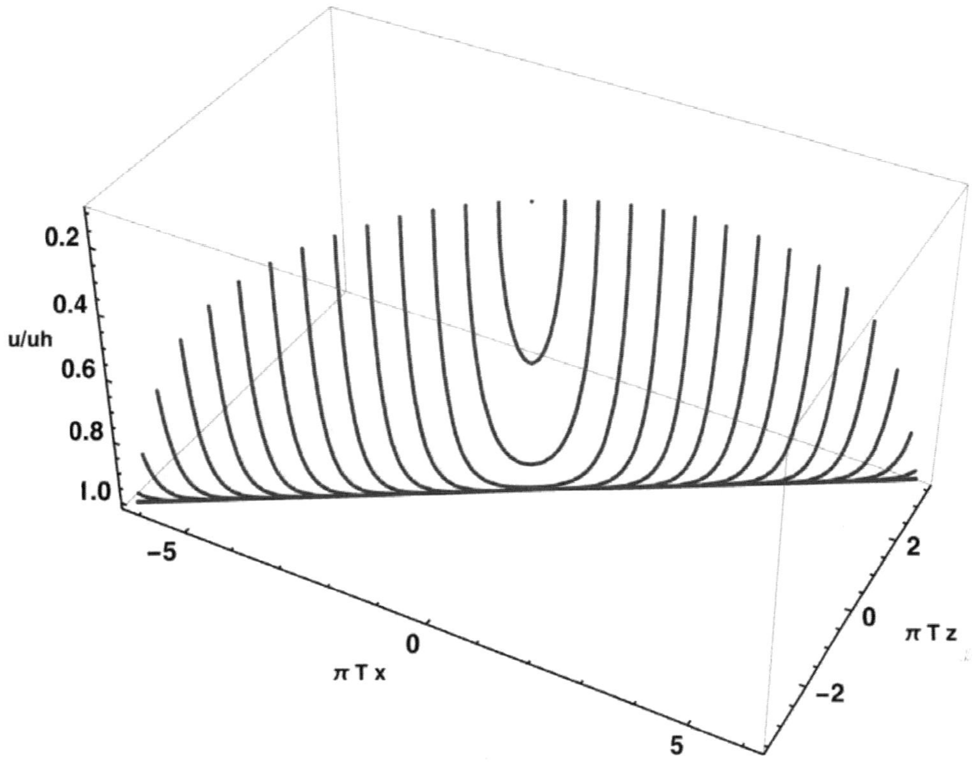

Figure 1. A typical falling string profile moving in the $x - z$ plane obtained numerically. The string is created at a point at $u_c = 0.1\, u_h$ and evolves to an extended object. The endpoints of the string move away from each other and fall toward the horizon.

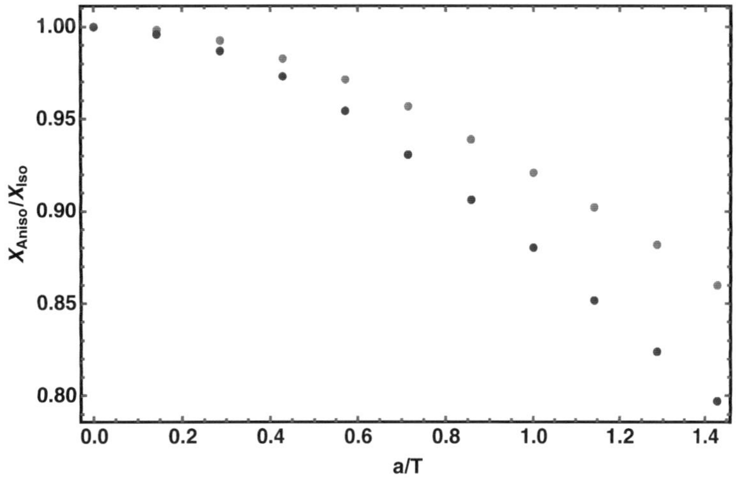

Figure 2. The ratio of the stopping distances in the anisotropic geometry to the $AdS - Sch$ metric for different values of a/T. Red dots show the ratio for $x_{Aniso} = x_{transverse}$, while the blue dots stand for $x_{Aniso} = x_{longitudinal}$. Here we fix the temperature of both backgrounds and the string has the same energy and virtuality in both backgrounds.

completely. It is useful to note that the virtuality of the corresponding jet in the field theory is defined as $Q^2 \equiv E_q^2 - P_q^2$, where P_q is the total spatial momentum of the quark. In order to have a better understanding of the jet quenching of a light probe in the anisotropic plasma we compare the stopping distance of a light quark moving in the anisotropic metric with the same string moving in the $AdS - Sch$ metric. To compare the anisotropic effect, we fix the temperature of the both backgrounds and the string moving in both background has the same energy as well. We find that even at the small anisotropy, $a/T < 1$ the stopping distance is smaller in both the transverse and the longitudinal direction which indicates stronger jet quenching of light probes traveling through the anisotropic medium, see figure 2.

4. Discussion

In this paper, we have studied the effect of anisotropy on an energetic jet traveling in the QGP. We have considered a point-like initial condition string which moves through the anisotropic background with high temperature or small anisotropy. The equation of motion for the string has been solved numerically and is plotted in figure 1 for a typical string. We have assumed that the string moves in the $x - z$ plane with an angle $\phi = \pi/3$ with respect to the direction of beam, z. Our results show that the string is highly suppressed in the longitudinal direction. In order to compare the effect of anisotropy, we have calculated the maximum stopping distance of an energetic jet traveling in a strongly coupled anisotropic plasma with and without anisotropy, figure 2. We have seen that even at small anisotropic metric, the stopping distance decreasing for both longitudinal (blue dots) and transverse direction (red dots) with respect to the static isotropic medium. Although, the quenching is stronger along the longitudinal direction.

Acknowledgments R. Morad wishes to thank the SA-CERN Collaboration, the South African National Research Foundation (NRF), and the National Institute for Theoretical Physics (NITheP) for their generous support.

References

[1] Adams J *et al.* (STAR) 2005 *Nucl. Phys.* **A757** 102–183 (*Preprint* nucl-ex/0501009)
[2] Adcox K *et al.* (PHENIX) 2005 *Nucl. Phys.* **A757** 184–283 (*Preprint* nucl-ex/0410003)
[3] Arsene I *et al.* (BRAHMS) 2005 *Nucl. Phys.* **A757** 1–27 (*Preprint* nucl-ex/0410020)
[4] Back B B *et al.* 2005 *Nucl. Phys.* **A757** 28–101 (*Preprint* nucl-ex/0410022)
[5] Maldacena J M 1999 *Int. J. Theor. Phys.* **38** 1113–1133 [Adv. Theor. Math. Phys.2,231(1998)] (*Preprint* hep-th/9711200)
[6] Witten E 1998 *Adv. Theor. Math. Phys.* **2** 253–291 (*Preprint* hep-th/9802150)
[7] Gubser S S, Klebanov I R and Polyakov A M 1998 *Phys. Lett.* **B428** 105–114 (*Preprint* hep-th/9802109)
[8] Aharony O, Gubser S S, Maldacena J M, Ooguri H and Oz Y 2000 *Phys. Rept.* **323** 183–386 (*Preprint* hep-th/9905111)
[9] Casalderrey-Solana J, Liu H, Mateos D, Rajagopal K and Wiedemann U A 2011 (*Preprint* 1101.0618)
[10] Florkowski W 2008 *Phys. Lett.* **B668** 32–35 (*Preprint* 0806.2268)
[11] Florkowski W and Ryblewski R 2009 *Acta Phys. Polon.* **B40** 2843–2863 (*Preprint* 0901.4653)
[12] Ryblewski R and Florkowski W 2008 *Phys. Rev.* **C77** 064906 (*Preprint* 0804.2427)
[13] Romatschke P and Strickland M 2003 *Phys. Rev.* **D68** 036004 (*Preprint* hep-ph/0304092)
[14] Arnold P B, Moore G D and Yaffe L G 2005 *Phys. Rev.* **D72** 054003 (*Preprint* hep-ph/0505212)
[15] Rebhan A, Strickland M and Attems M 2008 *Phys. Rev.* **D78** 045023 (*Preprint* 0802.1714)
[16] Mateos D and Trancanelli D 2011 *Phys. Rev. Lett.* **107** 101601 (*Preprint* 1105.3472)
[17] Janik R A and Witaszczyk P 2008 *JHEP* **09** 026 (*Preprint* 0806.2141)
[18] Chernicoff M, Fernandez D, Mateos D and Trancanelli D 2012 *JHEP* **08** 100 (*Preprint* 1202.3696)
[19] Giataganas D 2012 *JHEP* **07** 031 (*Preprint* 1202.4436)
[20] Chernicoff M, Fernandez D, Mateos D and Trancanelli D 2012 *JHEP* **08** 041 (*Preprint* 1203.0561)
[21] Armesto N, Salgado C A and Wiedemann U A 2004 *Phys. Rev. Lett.* **93** 242301 (*Preprint* hep-ph/0405301)
[22] Armesto N, Salgado C A and Wiedemann U A 2005 *Phys. Rev.* **C72** 064910 (*Preprint* hep-ph/0411341)
[23] Herzog C P, Karch A, Kovtun P, Kozcaz C and Yaffe L G 2006 *JHEP* **07** 013 (*Preprint* hep-th/0605158)
[24] Chesler P M, Jensen K and Karch A 2009 *Phys. Rev.* **D79** 025021 (*Preprint* 0804.3110)
[25] Chesler P M, Jensen K, Karch A and Yaffe L G 2009 *Phys. Rev.* **D79** 125015 (*Preprint* 0810.1985)
[26] Lawrence A E and Martinec E J 1994 *Phys. Rev.* **D50** 2680–2691 (*Preprint* hep-th/9312127)

A distinguishing observable of strongly- and weakly coupled plasmas: NLO momentum correlations of $b\bar{b}$ pairs in heavy ion collisions at $\sqrt{s} = 2.76\text{TeV}$

R Hambrock and WA Horowitz

Department of Physics, University of Cape Town, Private Bag X3, Rondebosch 7701, South Africa

E-mail: roberthambrock@gmail.com, wa.horowitz@uct.ac.za

Abstract. We use an energy loss model sensitive to thermal fluctuations [1] to compute the azimuthal and momentum correlations of $b\bar{b}$ pairs traversing a strongly coupled plasma from Pb+Pb collisions at LHC ($\sqrt{s} = 2.76\text{TeV}$). The azimuthal correlations are compared with those from perturbative QCD based simulations [2]. When restricted to leading order production processes, we find that the strongly coupled correlations of high transverse momentum pairs ($> 4\text{GeV}$) are broadened less efficiently than the corresponding weak coupling based correlations, while low transverse momentum pairs ($1 - 4\text{GeV}$) are broaded with similar efficiency, but with an order of magnitude more particles ending up in this momentum class. The strong coupling momentum correlations we compute account for initial correlations and reveal that the particle pairs suppressed from initially high momenta to the low momentum domain do not suffice to explain the stark difference to the weak coupling results in momentum correlations for $1 - 4\text{GeV}$. From this, we conclude that heavy quark pairs are more likely to stay correlated in momentum when propagating through a strongly coupled plasma than a weakly coupled one.

Keywords: Quark-Gluon Plasma, AdS/CFT Correspondence, Heavy Quarks

1 Introduction

The quark gluon plasma is of great interest since it represents our first case study of the emergent physics of the non-abelian gauge theory QCD. A key step in understanding this state of matter is identifying its relevant coupling strength. The perturbative techniques of QCD are only adequate in a weakly coupled plasma, with calculations for strongly coupled plasmas constrained to methods like AdS/CFT-based approaches or Resonance Scattering [3]. Both weak and strong coupling based approaches have had their respective successes in the past. For instance, experimental R^{π}_{AA} measurements show surprisingly consistent agreement with predicions from pQCD based models [4], while AdS/CFT based calculations have fared strongly by predicting a global lower bound on the shear viscosity-to-entropy ratio of QGP-like systems of $\frac{\eta}{s} \sim 0.1$ [5], when taking natural units, which is in line with hydrodynamic inferences from collider data at LHC and at RHIC [6].

IOP Conf. Series: Journal of Physics: Conf. Series **889** (2017) 012015 doi:10.1088/1742-6596/889/1/012015

Both frameworks show qualitative agreement with R_{AA}^{D} [4], suggesting they are attaining sufficient maturity to investigate more differential observables. We will argue that the momentum correlations of heavy quarks constitute a promising candidate as a differentiator between weakly and strongly coupled plasmas.

In [2], the azimuthal correlations of heavy $q\bar{q}$ pairs in a weakly coupled plasma in Pb+Pb collisions ($\sqrt{s}$ = 2.76TeV) were studied, both for a model involving purely collisional energy loss and one additionally incorporating radiative corrections. These weak coupling based azimuthal correlations provide a secondary indicator for the momentum correlations and we will compare them with computations from an AdS/CFT correspondence exploiting energy loss model sensitive to thermal fluctuations, the latter already having been introduced in [1]. As in [1], we will probe the spectrum of plausible [7] AdS/CFT based energy loss models with two plausible 't Hooft coupling constants ($\lambda_1 = 5.5$ and $\lambda_2 = 12\pi\alpha_s \approx 11.3$ with $\alpha_s = 0.3$) where for the former, temperature and the Yang-Mills coupling are equated, while for the latter constant, energy density and the coupling are equated.

The calculations will be performed for the same transverse momentum classes as in [2] and also both with leading order and next-to-leading order production processes used for the initialisation. Additionally, we will consider momentum correlations that take initial momentum correlations into account. These will provide evidence that heavy quarks traversing a strongly coupled plasma are more likely to stay correlated in momentum than they would if inside a weakly coupled plasma.

2 Energy Loss Model

2.1 Overview

The following will outline our computational procedure and its background. Subsequent to initializing the momenta of bottom quark pairs either to leading order with FONLL [8] or to next-to-leading order with aMC@NLO [9] using Herwig++ [10] for the showering, the production points of the bottom quarks are weigthed by the Glauber binary distribution [1]. The particles are propagated through the plasma via the energy loss mechanism described in 2.2 until the temperature in their local fluid cell drops below the Tc threshold and hadronization was presumed to occur or 8.6fm had passed, being the maximum time of the VISHNU background [6]. Finally, the bottom quarks are binned pairwise according to their relative azimuthal angle and each particle's final three-momentum.

2.2 Langevin Energy Loss

The stochastic equation of motion for a heavy quark in the fluid's rest frame is [11]

$$\frac{dp_i}{dt} = -\mu p_i + F_i^L + F_i^T \tag{1}$$

where F_i^L and F_i^T are longitudinal and transverse momentum kicks with respect to the quark's direction of propagation and with μ, the drag loss coefficient, being given by $\mu = \pi\sqrt{\lambda}T^2/2M_Q$ [12] where M_Q is the mass of a heavy quark in a plasma of temperature T with 't Hooft coupling constant λ. The correlations of momentum kicks are given by

$$\langle F_i^T(t_1)F_j^T(t_2)\rangle = \kappa_T(\delta ij - \frac{\vec{p}_i\vec{p}_j}{|p|^2})g(t_2 - t_1) \tag{2}$$

$$\langle F_i^L(t_1)F_j^L(t_2)\rangle = \kappa_L\frac{p_ip_j}{|p|^2}g(t_2 - t_1) \tag{3}$$

$\boldsymbol{\lambda_1 = 5.5}$

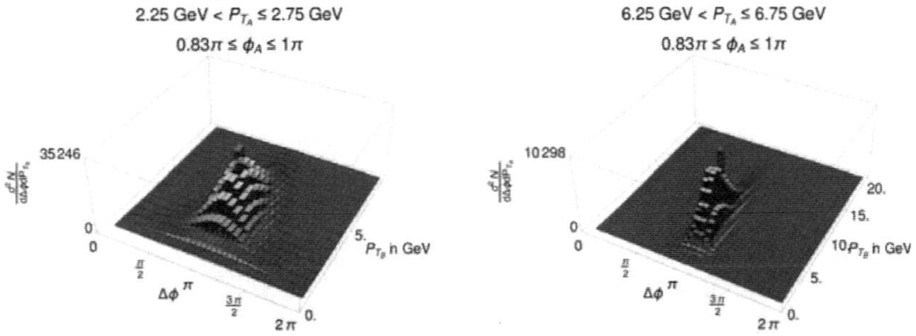

Figure 1: $\frac{d^2N}{d\phi dp_T}$ of $b\bar{b}$ pairs for $p_A = \{2.5, 6.5\}$GeV at 40-50 centrality

where g is only known numerically [1] and with

$$\kappa_T = \pi\sqrt{\lambda}T^3\gamma^{1/2} \tag{4}$$

$$\kappa_L = \gamma^2\kappa_T = \pi\sqrt{\lambda}T^{5/2} \tag{5}$$

As via Eqn. 5, the coupling of the longitudinal fluctuations to velocity grows as $\gamma^{5/2}$, thus growing significant extremely quickly [1]. For calculations without this scaling of the longitudinal fluctuations, see [13]. As reasoned in [1], the fluctuations are thus important to include for finite $\lambda \sim \mathcal{O}(10)$, as is the case for our probing models, where $\gamma_{crit}^{fluc} = \frac{M_Q^2}{4T^2}$ is lower than the speed limit on a quark, $\gamma_{crit}^{sl} = (1+\frac{2M_Q}{\sqrt{\lambda}T})^2$ where $g(0) = 1$, since any kick will be fully correlated to itself. If the time scale of momentum kick correlations is small compared the time scale determined by the drag coefficient, we can model the colouring as white noise, hence treat g as a Dirac delta [1]. By virtue of our requirement $\gamma < \gamma_{crit}^{sl}$, it follows that $t_{corr}\mu \sim \frac{1}{2}\sqrt{\lambda}\sqrt{\gamma}\frac{T}{M_Q} < \frac{1}{2}\sqrt{\lambda}\sqrt{\frac{4M_Q^2}{\lambda T^2}\frac{T}{M_Q}} = 1$, and we may thus safely approximate the colouring as white noise.

3 Leading Order Correlations

3.1 2D correlations

In Fig. 1 and Fig. 2, the $\frac{d^2N}{d\phi dp_T}$ correlations are depicted for representative sections of the respective p_T classes. We observe that, for low p_T, we attain very efficient broadening of the angular correlations. For mid p_T, the angular correlations are much tighter, however with greater broadening of the momentum correlations, at least in absolute terms. For $\lambda_2 = 11.3$, both angular and momentum correlations are much weaker than for $\lambda_1 = 5.5$, given the larger consequent drag coefficient of the former.

3.2 Azimuthal correlations

In [2], at leading order, the weak coupling based computations exhibited very efficient broadening of initial azimuthal correlations for low p_T $b\bar{b}$ pairs ([4 − 10]GeV), which were washed out once NLO production processes were taken into consideration.

Both for mid- and high-p_T ([4 − 10]GeV and [10 − 20]GeV respectively), the initial correlations survive to a large degree, both at leading order and at next-to-leading order, suggesting that they may still be observable in an experimental context.

$\lambda_2 = 11.3$

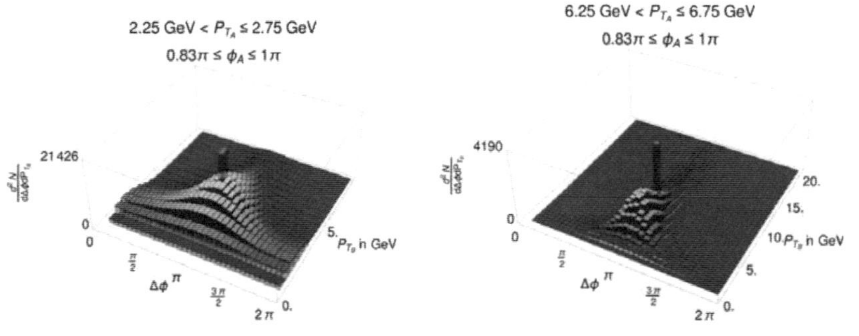

Figure 2: $\frac{d^2N}{d\phi dp_T}$ correlations of $b\bar{b}$ pairs for $p_A = \{2.5, 6.5\}$GeV at 40-50 centrality

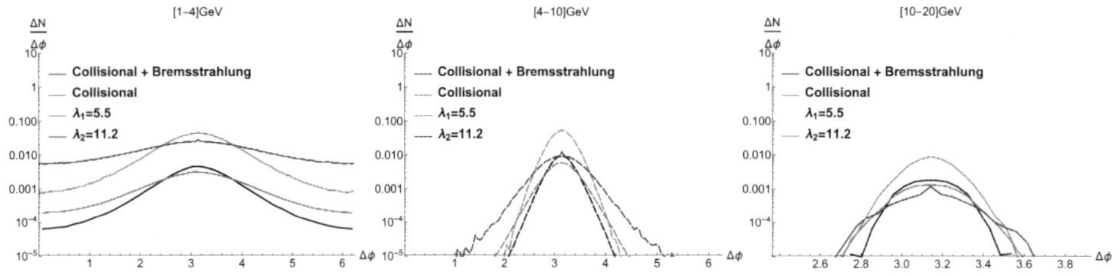

Figure 3: $\frac{dN}{d\phi}$ correlations for the specified classes.

We compare our strong coupling azimuthal correlations to the weak coupling ones in Fig. 3. For $[10-20\text{GeV}]$, our correlations are significant more peaked at their initial back-to-back correspondence. At $[4-10\text{GeV}]$, this observation still holds for the upper bound of our parameters with $\lambda_1 = 5.5$, while the $\lambda_2 = 11.3$ bounded result is of similar magnitude but looser angular correlation than either the collisional or the collisional + Bremsstrahlung based results. In the $[1-4\text{GeV}]$ range, the azimuthal correlations are almost entirely washed out for $\lambda_2 = 11.3$, while for $\lambda_1 = 5.5$, they are broadened with similiar efficiency to the weak coupling results.

3.3 Momentum correlations

What is striking about the low p_T correlations ($[1-4]$GeV) depicted in Fig. 3 is the order of magnitude difference between strong and weak coupling.

Naïvely, one may expect this to be caused by a more efficient suppression of high momentum particles in a strongly coupled plasma than a weakly coupled one. If we take initial momentum correlations into account Fig. 4, we find that the contribution of particles initially in a higher momentum class being suppressed down to $[1-4]$GeV clearly do not suffice to account for the order of magnitude difference observed. In fact, the dominant portion of particles with low final transverse momentum had low transverse momentum initially too. On the one hand, this could be due to these particles ending up in other final momentum classes. From Fig. 3, it is clear they

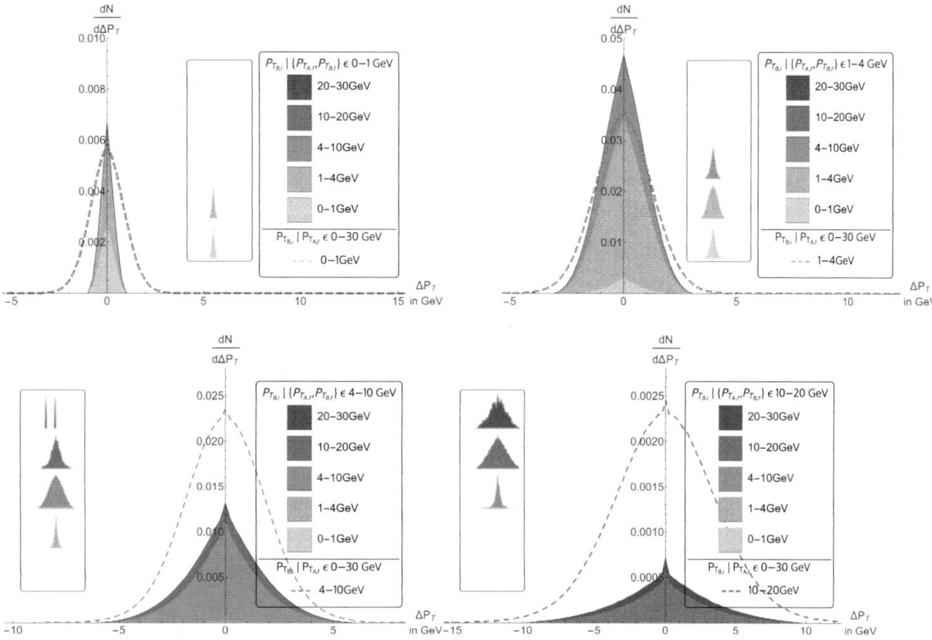

Figure 4: $\frac{dN}{d\Delta p_T}$ of $b\bar{b}$ pairs with initial momentum in some given class.

do not end up in [4 − 10]GeV or [10 − 20]GeV. This only leaves [0 − 1]GeV and [20 − ∞]GeV, the latter being highly unlikely. A more plausible explanation is that in a weakly coupled plasma, $b\bar{b}$ pairs are much more weakly correlated in momentum and are thus more likely to end up in distinct momentum classes.

Since this effect is observed in the low momentum domain, one may postulate that the momentum fluctuations in a weakly coupled plasma are more significant that in a strongly coupled plasma. Thus, it would be more likely for low momentum heavy quarks to receive momentum kicks that elevate them to higher momentum classes in the weakly coupled plasma.

4 Next-to-leading Order Correlations

4.1 Azimuthal correlations

As depicted in Fig. 5, the correlations for the [1 − 4]GeV class are entirely washed out, just as observed for the pQCD calculations in [2]. For mid- to high-p_T, while the peak around $\Delta\phi = \pi$ has been broadened, the computations suggest that this signal should still be observable in an experimental context. Unfortunately, we had no data from weak coupling calculations for a similar centrality to compare with.

5 Conclusion & Outlook

We have compared the azimuthal correlations predicted by pQCD and AdS/CFT based computations and found that, while the azimuthal correlations are qualitatively similar, the momentum correlations tell a different tale. In particular, the surprise of our findings is the large dissimilarity in low momentum correlations of the pQCD and

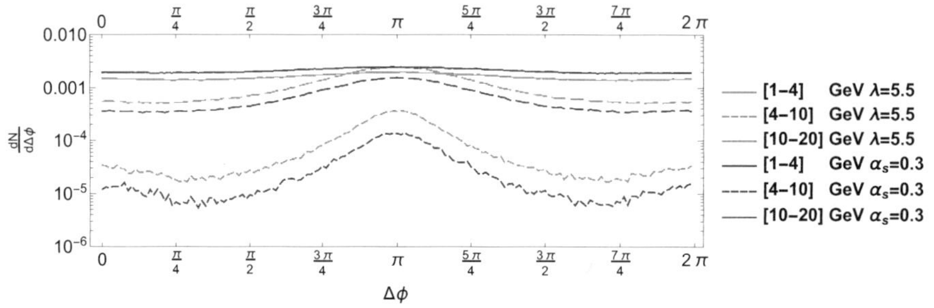

Figure 5: Final $\frac{dN}{d\phi}$ correlations for NLO initialization.

AdS/CFT based simulations. Thus, bottom quark momentum correlations present an opportunity to distinguish between the energy loss mechanisms of the two frameworks.

Although stronger momentum fluctuations in the weakly coupled plasma have been identified as a plausible explanation for this disparity, it can only be verified by computing initial correlations for the weak coupling based correlations as well.

Furthermore, whether this order of magnitude difference in predictions for low p_T correlations of heavy quarks exposes weaknesses in either or both of the frameworks cannot be declared until experimental data of bottom quark momentum correlations emerge. Strong coupling based approaches have fared better in the low momentum domain, where pQCD is restrained by uncertainties in the running coupling.

While discriminating initial momenta in the momentum correlation plots Fig. 4 is illuminating, producing these is impossible in the current paradigm of detectors, and thus their applicability is restricted to simulations. That being said, the integrated momentum correlations may constitute a potent experimental discriminator of strongly and weakly coupled plasmas and demand further investigation.

References

[1] Horowitz W A 2015 *Phys. Rev.* **D91** 085019 (*Preprint* 1501.04693)
[2] Nahrgang M, Aichelin J, Gossiaux P B and Werner K 2014 *Phys. Rev.* **C90** 024907 (*Preprint* 1305.3823)
[3] Casalderrey-Solana J, Liu H, Mateos D, Rajagopal K and Wiedemann U A 2011 (*Preprint* 1101.0618)
[4] Horowitz W A 2013 *Nucl. Phys.* **A904-905** 186c–193c (*Preprint* 1210.8330)
[5] Gubser S S, Klebanov I R and Peet A W 1996 *Phys. Rev.* **D54** 3915–3919 (*Preprint* hep-th/9602135)
[6] Shen C, Heinz U, Huovinen P and Song H 2011 *Phys. Rev.* **C84** 044903 (*Preprint* 1105.3226)
[7] Gubser S S 2008 *Nucl. Phys.* **B790** 175–199 (*Preprint* hep-th/0612143)
[8] Cacciari M, Frixione S, Houdeau N, Mangano M L, Nason P and Ridolfi G 2012 *JHEP* **10** 137 (*Preprint* 1205.6344)
[9] Alwall J, Frederix R, Frixione S, Hirschi V, Maltoni F, Mattelaer O, Shao H S, Stelzer T, Torrielli P and Zaro M 2014 *JHEP* **07** 079 (*Preprint* 1405.0301)
[10] Bahr M *et al.* 2008 *Eur. Phys. J.* **C58** 639–707 (*Preprint* 0803.0883)
[11] Moore G D and Teaney D 2005 *Phys. Rev.* **C71** 064904 (*Preprint* hep-ph/0412346)
[12] Herzog C P, Karch A, Kovtun P, Kozcaz C and Yaffe L G 2006 *JHEP* **07** 013 (*Preprint* hep-th/0605158)
[13] Hambrock R and Horowitz W A 2017 AdS/CFT predictions for azimuthal and momentum correlations of $b\bar{b}$ pairs in heavy ion collisions *8th Int. Conf. on Hard and Electromagnetic Probes of High-energy Nuclear Collisions: Hard Probes 2016 (HP2016) Wuhan, Hubei, China, September 23-27, 2016* (*Preprint* 1703.05845)

IOP Conf. Series: Journal of Physics: Conf. Series **889** (2017) 012016 doi:10.1088/1742-6596/889/1/012016

System Control Applications of Low-Power Radio Frequency Devices

Roger van Rensburg

School of Physics, University of the Witwatersrand, Johannesburg 2050, South Africa

E-mail: roger.vanrensburg@wits.ac.za

March 2017

Abstract. This paper conceptualizes a low-power wireless sensor network design for application employment to reduce theft of portable computer devices used in educational institutions today. The aim of this study is to design and develop a reliable and robust wireless network that can eradicate accessibility of a device's human interface. An embedded system supplied by an energy harvesting source, installed on the portable computer device, may represent one of multiple slave nodes which request regular updates from a standalone master station. A portable computer device which is operated in an undesignated area or in a field perimeter where master to slave communication is restricted, indicating a possible theft scenario, will initiate a shutdown of its operating system and render the device unusable. Consequently, an algorithm in the device firmware may ensure the necessary steps are executed to track the device, irrespective whether the device is enabled. Design outcomes thus far indicate that a wireless network using low-power embedded hardware, is feasible for anti-theft applications. By incorporating one of the latest Bluetooth low-energy, ANT+, ZigBee or Thread wireless technologies, an anti-theft system may be implemented that has the potential to reduce major portable computer device theft in institutions of digitized learning.

Keywords: micro-controller unit (MCU), portable computer device (PCD), Bluetooth low-energy (BLE), radio frequency (RF), system-on-chip (SoC)

1. Introduction

In 2015 the South African (SA) government spent billions of rands in modernizing learning institutions by providing tablets to schools. The government stated that their goal for educational learning in SA is to digitize schools by the year 2020. Unfortunately, it was reported that more than 3000 out of 88000 tablets provided, were lost due to theft [1]. The large project undertaking was in jeopardy and under a substantial amount of stress because of the significant financial loss incurred. Consequently, more than 88000 tablets were recalled and fitted with anti-theft technology. The upgrade of the devices enabled the government to involve investigative authorities in the tracking of stolen tablets [2]. However, the tracking systems appeared only operational with the device powered on to communicate with Global Positioning Satellites (GPS) in outdoor environments.

With deficits present currently in the protection of tablets, there is need to augment the security of such devices by taking advantage of the wireless technologies today. For this reason, this research proposes a low-power wireless network design to protect PCD against theft. With such a system in place, theft may not be eliminated overnight but will make a stolen device difficult to use and less appealing to thieves. When the device functionality is removed that render the hardware of the user interfaces useless outside the working perimeter of the wireless network, a tracking system may be initiated to aid investigating authorities in the retrieval of the device.

The conceptual phase of this project will mainly consist of four stages. The first stage may deal with the design and development of a proposed wireless *master-slave* network topology followed by the electronics development of the embedded hardware. The second phase may consist of researching energy harvesting methods to power each of the *slave* nodes operating independently of the *device* power supply. The third phase may consist of developing firmware to render the *device* inoperable. Optionally, a final phase may be included, which requires a gateway in the wireless network that sends notifications to the learning institution when the *device* is rendered inoperable.

IOP Conf. Series: Journal of Physics: Conf. Series **889** (2017) 012016 doi:10.1088/1742-6596/889/1/012016

2. Methodology

2.1. Introduction

Embedded hardware that integrate wireless technologies and processing capabilities on a miniature SoC will be fully utilized in the design to realize a low-powered network topology. The problem statement requires that for such a network to be practical, network independence from an Internet Service Provider (ISP) to render the *device* unusable is needed. With network independence from carriers, the anti-theft system may be used in any geographic location. The wireless technology must also provide security that prevents unauthorized connections to the ad hoc network and offer reliable communications links over long distances, even when surrounded with other Industrial, Scientific and Medical (ISM) band frequencies. In the following section, the network topologies containing low-power RF technologies are proposed.

2.1.1. Star network topology: BLE wireless technology, also known as Bluetooth Smart, is a connection orientated communications standard. BLE has proven to be more power efficient that provide higher data throughput compared to other wireless technologies [3]. High cyclic redundancy checks are guaranteed which offers very robust and secure communications. Compared to ANT+ and ZigBee, BLE is fast becoming more progressively available in smart devices today.

A star network topology is firstly proposed which consist of a *master* station and multiple PCD. Each PCD, consist of two slave nodes which integrate a low-power BLE radio and an ARM Cortex M4 processor on a single chip [4]. As depicted by Fig. 1, the PCD in Fig. 1C must request regular data updates to the *master* in Fig. 1F. Similarly all other PCD in the network are setup identically. BLE only allows a single *slave* to be connected to one *master* at a time. For this reason, each PCD incorporates two *slave* nodes shown by Fig. 1D and denoted by MS_1 and DS_1, for connections to the *master* and *device*. The *master*, which can connect and communicate up to 20 *slaves*, will be located in a secure room and does not rely on low-power consumption requirements. This proposed BLE topology, however, will only be able to operate in a space over a limited distance where there is no physical obstructions of the wireless network.

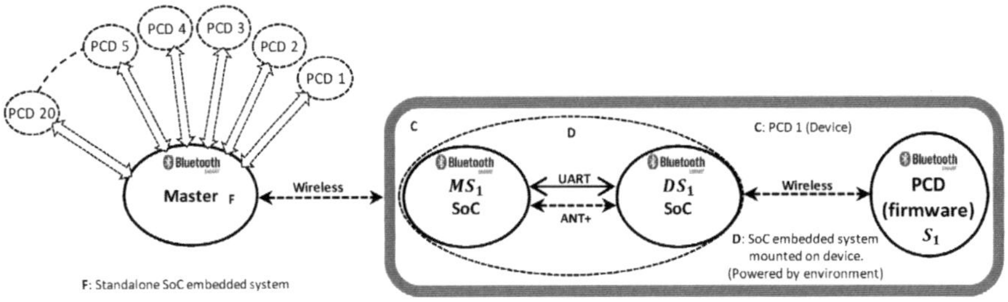

Figure 1: Proposed BLE star wireless network (concept design).

Each node (*master*, MS_1, DS_1 and PCD, incorporates a control algorithm to realise the communication system from PCD to *master* at the application layer of the network. The control algorithms, illustrated as flow-charts, in Fig. 2 are summarized as follows: The DS_1 node will advertise a connection to S_1, and similarly, the MS_1 node will advertise a connection to the *master*. Since BLE is a connection orientated wireless technology, only the *master* and PCD may initiate a connection to their respective slave node. The PCD must however run the BLE stack on its firmware while also acting as a *master* in the network. After DS_1 advertised a connection, the PCD will connect and bond with DS_1. Similarly, after MS_1 advertise a connection, the *master* will connect and establish a bond with the MS_1. Bonding of the *slaves* will ensure that whenever a connection is lost with the *master* or PCD, a secure connection with unique *slave* identification will be established to reconnect with DS_1 or MS_1 [5]. The PCD will start requesting regular data updates from DS_1 via BLE while DS_1 will request regular data updates from MS_1 via the Universal Asynchronous Receiver Transmitted (UART) or wireless ANT+ connection. Similarly, the *master* will update MS_1. If an acknowledgement error by the PCD is detected by a request not received by DS_1, after a certain number of attempts have been exceeded as illustrated by the control chart in Fig. 2B and 2C, the PCD will initiate its shut-down as illustrated by the control chart in Fig. 2A. Consequently, the *slaves* DS_1 or MS_1 will reset their MCU by their respective watchdog timers. Ideally, the initiation of the PCD shutdown may be accomplished by software development of the

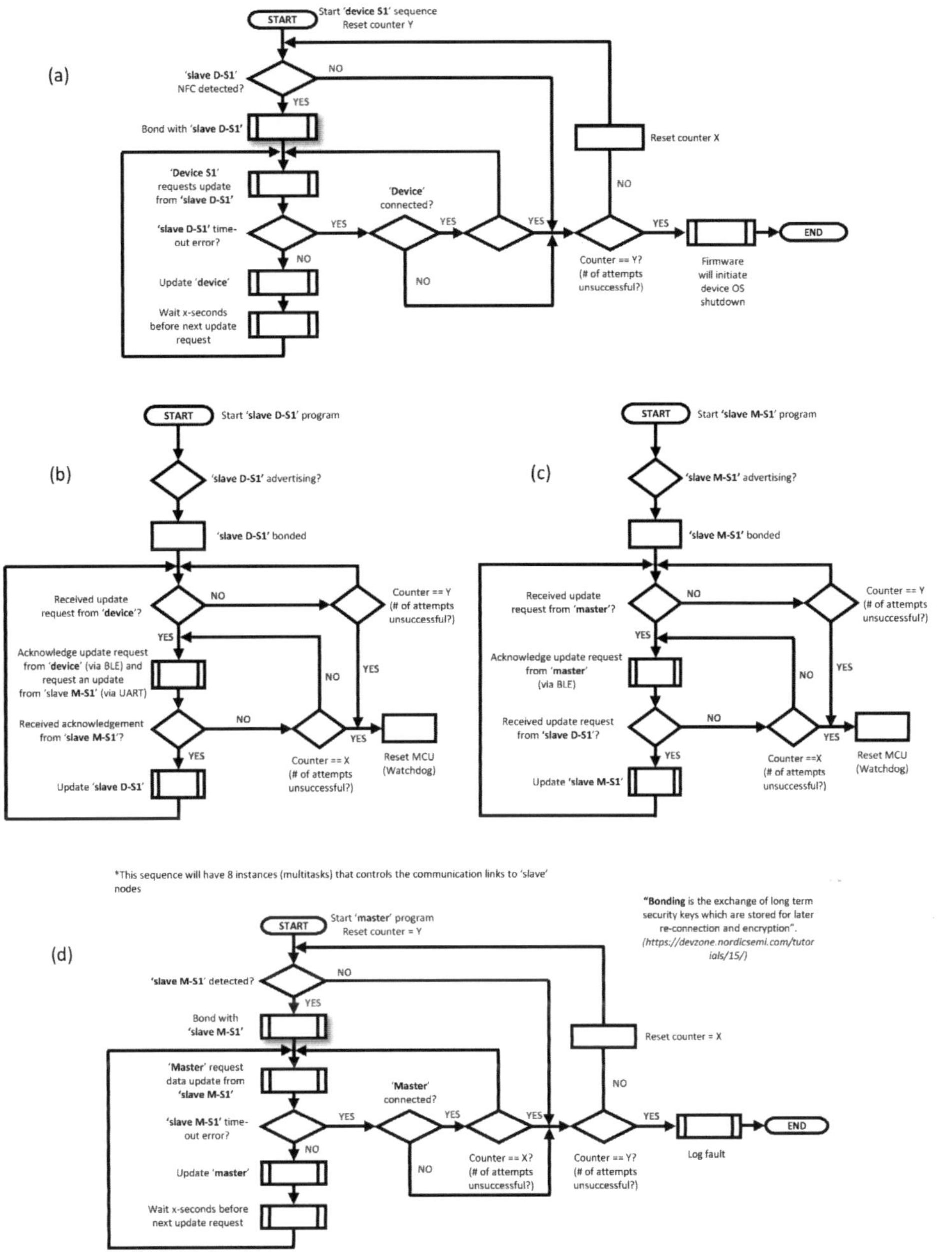

Figure 2: Proposed BLE network topology algorithms (concept design).

firmware. Potentially, after the PCD is rendered inoperable, the theft detecting mechanism may enable a tracking system. If the PCD is operated in an undesignated area or tampering is detected, the firmware must ensure that the device is rendered unusable. Subsequently, when the PCD detects a GPS, tracking must be enabled. If the connection to the GPS is lost, the state will continuously try to reconnect to enable the tracking system. For tracking, it is important that sufficient battery power is stored, before an OS shutdown, to execute the algorithm successfully.

2.1.2. Hybrid mesh-star network topology: The nRF52 from Nordic Semiconductors provide application programming interfaces which support the BLE, ANT+ and Thread stacks while other low-powered SoC is widely available for ZigBee such as the CC2640/CC2650 from Texas

Instruments [6]. The latest nRF52 low-powered SoC solution can run BLE and the proprietary 802.15.4 stacks concurrently. A developer may also choose to utilize ZigBee on the nRF52 SoC which requires the protocol to be built on top of the hardware layer of the IEEE 802.15.4 standard. Fig. 3 proposes a hybrid mesh-star topology (ANT+, ZigBee, or Thread) that offers a distinct advantage over the proposed BLE topology in Fig. 1 by what is known as mesh multi-hopping in a network. Furthermore, the ZigBee and Thread protocol mesh topologies, allows hundreds of nodes to communicate over the network and may be adapted to the proposed network topology.

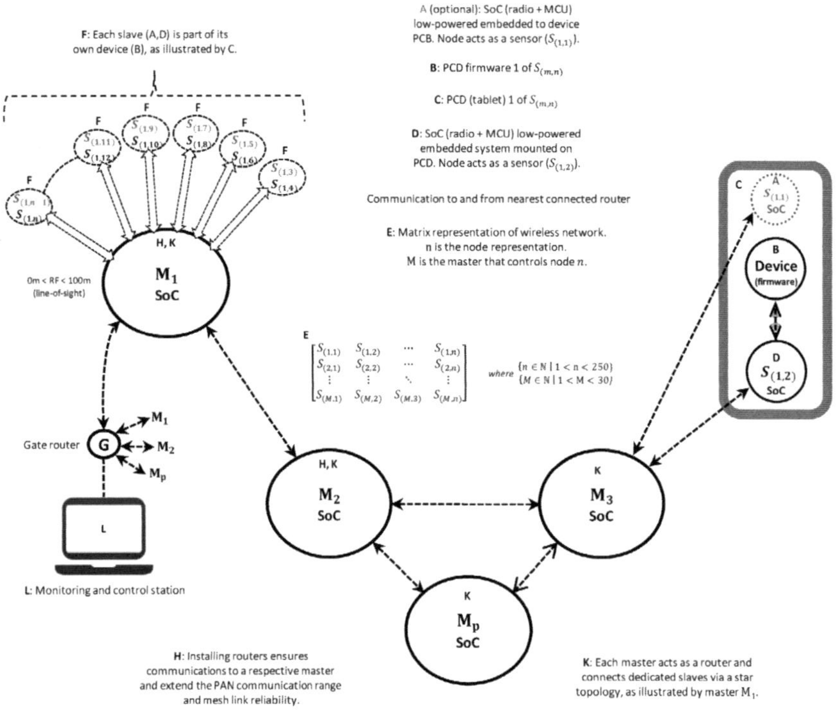

Figure 3: Proposed hybrid mesh-star wireless network (concept design).

In Fig. 3, a Wireless Sensor Network (WSN) is configured in a building on a campus (large educational institution) area. ZigBee and Thread uses Carrier Sense Multiple Access with Collision Avoidance (CSMA-CA) to ensure reliable and robust communications in the mesh network which provides highly secure AES-128 standard encryption and decryption of packets over the air [7]. A *master* acts as a router and can relay data packets from slaves or *end-devices* to other routers in the network. The router will continually run the SoC radio and processor program and does not rely on low-power consumption requirements. A PCD shown in Fig. 3C, shown as the *end-devices* $S_{(1,1)}$ and $S_{(1,2)}$, requests regular updates about connectivity from its *master* or from the nearest *router* in the network. *End-devices* consumes very little power since the SoC goes into sleep mode when not communicating with its *master*. However, the ZigBee and Thread wireless technologies, does not provide the routing of packets directly over the network by consecutive *end-devices* [8].

Optionally in the proposed design, if an acknowledgement is not received by $S_{(1,1)}$ from node M_3, then node $S_{(1,1)}$ will command the PCD shutdown by means of the integrated SoC hardware to the PCD. Ideally, if the PCD firmware in Fig. 1B, does not confirm from node $S_{(1,2)}$ its connectivity to the mesh network by node M_3, a shutdown of the PCD OS must be initiated. Node $S_{(1,2)}$ may be connected over the most reliable link to a router as illustrated by node M_3. If the PCD is operated in a building where it cannot directly communicate with node M_3, an alternative router have to be selected in the mesh network. Such autonomous *slave* to *master* link behavior requirements, make this application ideal for implementation for low-power Thread networks. Multi-hop hybrid mesh network topologies may be used in different building areas where PCD have to communicate over larger distances and obstacles are unavoidable. Additional routers may be installed to enable reliable communications between all masters in the network. Each node in the network will consist of the miniature nRF52 SoC shown in Fig. 4 and illustrated in Fig. 2A and 2D. As mentioned previously, the SoC in Fig. 2A may potentially be embedded in the tablet PCB hardware to render the device inoperable. This solution will disable the PCB, but will be further explored regarding its plausibility. More realistically, a firmware-based solution however, will render the PCD inoperable when no heartbeat is sensed from $S_{(1,2)}$ to the wireless mesh network.

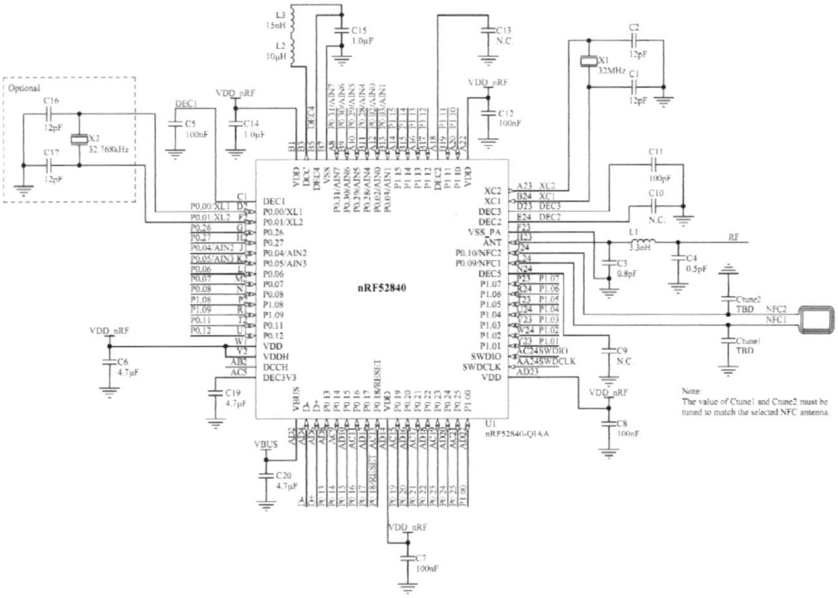

Figure 4: nRF52840 SoC IO schematic (7 mm x 7 mm 73 pin AQFN package) [6]

2.1.3. Energy harvesting: The wireless technologies described utilize low-power consumption devices that may make use of energy harvesting methods to convert energy from either or both RF and thermal sources to regulate a supply or charge a battery of a *slave* or *end-device* node. Harvesting energy from the environment requires Integrated Circuits (IC) with highly optimized DC-DC boost converters and power management circuitry capabilities. An experimental study will be conducted to determine the output power of a suitable harvesting source. Once these output levels are identified and found to be feasible, an energy harvesting solution may potentially be implemented that may integrate either the LTC3108 [9] or the MAX17710 [10] IC with the SoC. Thermal electric energy harvesters that utilize the Peltier element may offer higher source output levels compared to RF harvesting [11] because of small temperature differences that generate voltages. A PCD may naturally generate heat at its power source and central processor unit over a certain amount of working operation. A Peltier element may be installed on the PCD that absorbs the transferred heat from the junctions [12]. The electrical energy generated from the harvester can then charge a coin or thin cell battery which directly supply power to the SoC with the appropriate electronics designs in place. For this particular application, further research is necessary to determine feasibility for energy harvesting using RF or thermal sources.

3. DISCUSSION

The network topology proposed implementing BLE has shown to be more complex which may provide better network stability and security compared to other wireless technologies. At the moment, BLE has shortfalls by not allowing the mesh topology design shown in Fig. 3 and only providing a limited number of node connectivity to a *master* in the network. Mesh network topologies may be simulated in Matlab or NS-3 to test routing algorithms for implementation on embedded hardware, with a particular focus on ZigBee or Thread wireless technologies. By modeling the environment where the wireless network will be distributed, signal strength path loss exponents can be determined and routers can be placed optimally in a building area. With multi-hop formation capabilities in a network, greater RF communication distances are possible which extends routing distances for reliable data delivery from *device* to its *master*. The radio power consumption of the nRF52832, is

IOP Conf. Series: Journal of Physics: Conf. Series **889** (2017) 012016 doi:10.1088/1742-6596/889/1/012016

rated at four dBm (with wireless distances of up to 100 m line-of-sight) and estimated by the 'BLE online profiler' to only consume around 15 uW of power over two-second regulatory updates between *master* and *slave* [13]. The power consumption rating excludes the UART or ANT+ communication between DS_1 and M_1 shown in the proposed network topology of Fig. 1. The SoC will therefore consume more power than previously stated because of the utilization of the processor's UART peripheral or multi-protocol functionality of the radio with ANT+. Nonetheless, the radio power specifications indicate that a single *slave* node can run for more than four years on a coin cell battery rated at 300 mAh and efficiency of 70 %. With such low power constraints, *slave* nodes can be powered by thin cell batteries or energy harvesting sources. Similar low-power consumability of the *slave* nodes is possible in the proposed multi-hop mesh wireless topology, assuming a *slave* does not have a duty cycle (*device* active time over total time) greater than 1 % [8].

A typical solution to prevent flashing of a PCD firmware is for manufacturers to add authentication ability to their hardware [14]. Some companies provide solutions in preventing devices from factory reset and installation of new PCD firmware [15]. Such companies use patented technology whereby their firmware is built into the hardware during the PCD manufacturing process [15]. This study, however, will initially focus on an in-depth experimental study on the reliability of the proposed wireless mesh technology followed by a proprietary firmware-based solution to render the PCD unusable.

References

[1] Phaladi B 2015 http://citizen.co.za/news/news-national/382489/schools-tablet-theft-shock/

[2] Writer S 2015 https://businesstech.co.za/news/gauteng-withdraws-88000-tablets-from-schools-because-of-theft/

[3] Dementyev A, Hodges S, Taylor S and Smith J 2013 Power consumption analysis of bluetooth low energy, zigbee and ant sensor nodes in a cyclic sleep scenario *2013 IEEE International Wireless Symposium (IWS)* pp 1–4

[4] Nordic-Semiconductor 2016 https://www.nordicsemi.com/Products/nRF52-Series-SoC/

[5] Heydon R 2007 *Bluetooth Low Energy: The Developers Handbook* 1st ed (Indiana: Prentice Hall)

[6] Nordic-Semiconductor 2017 https://infocenter.nordicsemi.com/index.jsp

[7] Gislason D 2008 *ZigBee Wireless Networking* 1st ed (Burlington: Newnes)

[8] Farahani S 2008 *ZigBee Wireless Networks and Transceivers* 1st ed (Burlington: Newnes)

[9] Linear-Technology 2017 http://www.linear.com/product/LTC3108

[10] Maxim-Integrated 2017 https://www.maximintegrated.com/en/products/power/MAX17710.html

[11] Gudan K, Chemishkian S, Hull J J, Thomas S J, Ensworth J and Reynolds M S 2014 A 2.4ghz ambient rf energy harvesting system with -20dbm minimum input power and nimh battery storage *2014 IEEE RFID Technology and Applications Conference (RFID-TA)* pp 7–12

[12] Nesarajah M and Frey G 2016 Thermoelectric power generation: Peltier element versus thermoelectric generator *IECON 2016 - 42nd Annual Conference of the IEEE Industrial Electronics Society* pp 4252–4257

[13] Nordic-Semiconductor 2016 https://devzone.nordicsemi.com/power/

[14] Zetter K 2015 https://www.wired.com/2015/02/firmware-vulnerable-hacking-can-done/

[15] Absolute 2016 https://lojack.absolute.com/en/persistent/

IOP Conf. Series: Journal of Physics: Conf. Series **889** (2017) 012017 doi:10.1088/1742-6596/889/1/012017

A Characterisation of the ATLAS ITk High Rapidity Modules in AllPix and EUTelescope

Ryan Justin Atkin

Department of Physics, University of Cape Town, Private Bag X3, Rondebosch 7701, Cape Town, South Africa

E-mail: rjatkin93@gmail.com

Abstract. The upgrade of the LHC to the high luminosity LHC (HL-LHC) will result in far more collisions occurring per bunch crossing, in turn producing more particles per second. Consequently, the current detectors will need to be upgraded to accommodate the large increase in radiation and data acquisition as well as a need to improve the tracking efficiency for the high pile-up environment. One of the main upgrades to the ATLAS detector is the complete overhaul of the inner detector (ID) by replacing it with an all silicon Inner Tracker (ITk). A simulation of the ITk will be required for performance predictions as well as for testing sample sensors in testbeams. The current testbeam software of Allpix and EUTelescope are written completely using Cartesian definitions, however some of the geometries in the ITk have radial definitions. In particular, the R0 geometry of the strip end-cap is in need of a radial description. Presented is the work behind creating a radial geometry for the R0 module in Allpix (using Geant4 descriptions) and EUTelescope (using TGeo descriptions).

1. Introduction

With the discovery of the Higgs boson in 2012[1][2] at the Large Hadron Collider (LHC) and with new searches pushing the limits of the current LHC and its detectors, an upgrade of the LHC to the High Luminosity LHC (HL-LHC) has been planned. This upgrade will occur during the third long shutdown (LS3) for 30 months from the beginning of 2024 till half way through 2026[3]. The upgrade will increase the instantaneous Luminosity to an ultimate value of $\mathscr{L}_{ins} = 75\text{nb}^{-1}.\text{s}^{-1}$[3], around 7.5 times the current instantaneous Luminosity. This will result in a total integrated luminosity of over $\mathscr{L} = 3000\text{fb}^{-1}$ during the 10 years of operation and up to an average of $\mu = 200$ collisions per bunch crossing. These improvements will greatly increase the statistics available for analysis while at the same time exceeding the current detectors' design capabilities with respect to pile-up management and radiation tolerance. Therefore the detectors will require an upgrade themselves. In particular, the ATLAS detector's main upgrades (phase-2 upgrades) will occur during LS3 as laid out in the Letter of Intent (LoI)[4]. The focus will be on upgrading the current inner detector (ID) to the inner tracker (ITk) as well as the upgrade of the Trigger and Data AcQuisition systems (TDAQ). Other areas include the forward calorimeters, muon spectrometer and computing and software. The purpose of the ATLAS detector during the HL-LHC will be on precision Higgs measurements, Vector Boson Fusion (VBF) and Scattering (VBS), as well as searches for new physics.

1.1. ITk

The largest contribution to the ATLAS phase-2 upgrade will be the complete overhauling of the current ID, coming in at around 50% of the total upgrade cost. A comparison of the current ID to the future ITk is shown in Fig. 1. The Transition Radiation Tracker (TRT) will not be present in the ITk as this type of detector will become saturated in the high pile-up environment of the HL-LHC and thus make it incapable of precision tracking. This removal will make the ITk a full silicon semiconductor tracker divided into the strip detector, elongated sensors capable of 1D space point detection, and the pixel detector, square sensors capable of 2D space point detection. The pseudorapidity range will also be increased to $|\eta| < 4$ while having less inactive material in the tracking volume. The higher rapidity range will be important for the VBF, VBS and new physics searches, as well as improving E_T^{miss} resolution and pile-up jet rejection[3].
There are currently two designs for the ITk; the extended layout and the inclined layout, where the extended layout can be seen in Fig. 1. The inclined layout will be exactly the same as the extended except that the pixel barrel in the range $|\eta| \gtrsim 1$ will be inclined towards the interaction point. The extended will be less challenging to build and will allow for long pixel clusters while the the inclined will have less silicon for the particles traverse and will get more hits in the forward region[6].
The main reason for a completely new inner tracker is that the luminosity during HL-LHC will be over 7 times the design value of the current ID. The ID was designed to deal with an average of 23 p-p collisions per bunch crossing[3], not the proposed 200 during HL-LHC and the current pixel detectors were only intended to deal with radiation from 400fb^{-1} of data[3]. The current granularity of the the ID as well would fail at pattern recognition and have a poor track finding efficiency and so needs to be improved.

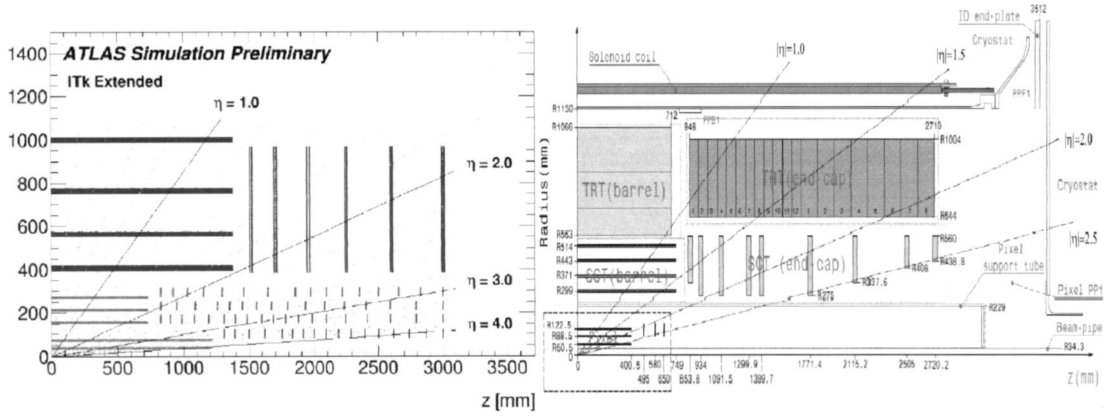

Figure 1. Diagram comparing the new Inner Tracker (left) to the current Inner Detector (right)[3][5]. On the left, the blue and the red are respectively the strip and pixel semiconductor detectors with the horizontal lines being the barrels and the vertical lines being the endcaps. As can be seen, there will be no transition radiation based detector in the ITk and the pseudo-rapidity, η, range will be increased to 4. The layout for the ITk is that of the extended layout.

2. R0 module

The strip endcap local support structure, known as a petal, will be installed castellated with 32 petals per endcap disk[3]. The petal will have 6 rings of modules on both faces, one in each of the first 3 rings and two in each of the outer 3 rings. They will be arranged from the R0

IOP Conf. Series: Journal of Physics: Conf. Series **889** (2017) 012017 doi:10.1088/1742-6596/889/1/012017

Table 1. R0 specifics

Row number	nStrips	nChips	Inner Radius [mm]	Length [mm]	Min/Max pitch [μm]	Angular Pitch [μrad]
0	1026	8	384.5	19	73.5/84	193.3
1			403.5	24		
2	1154	9	427.5	28.9		171.8
3			456.4	32		

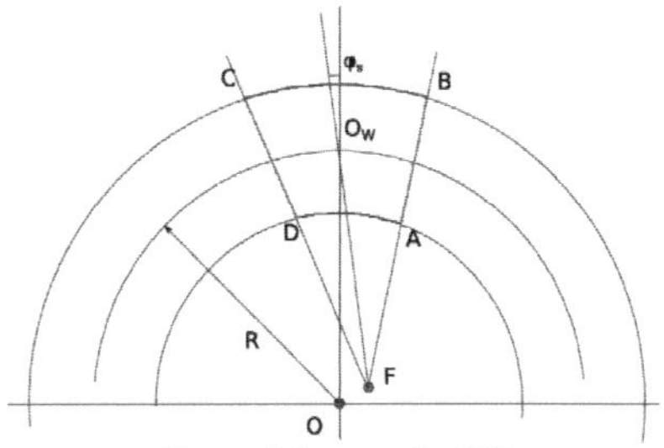

Figure 2. Diagram illustrating how the R0 module is defined[7]. The strips will be parallel to the sides in red, providing a stereo angle $\phi_s = 20$mrad. The radius of the sensor center O_W will be at R=438.6mm.

module closest to the beam pipe to the R5 modules furthest from the beam pipe. A module is a compound device comprising of a power board and one or two hybrids glued to a silicon semiconductor sensor, where the hybrid has ABC130 readout chips (ATLAS Binary Chip at 130nm thick) and hybrid control chips glued to a kapton board.

The shape of the R0 sensor is known as a stereo annulus and a simple definition of this geometry is shown in Fig. 2. The inner and outer curved edges will be rings concentric to the center of the pipe while the straight sides will converge to a point offset from the beam centre. The strips will be placed parallel to these offset sides and focus on the same offset point, providing a stereo angle of 20mrad built into the sensor and a total of 40mrad[3] when combined in conjunction with the sensor on the other side of the petal. The stereo angle was chosen to be built into the sensor as rotating the sensor during installation to obtain the stereo angle would have complicated the installation process. As the strips are only capable of 1D space point measurements, the strips are placed at a stereo angle which when combined with the strips on the other side of the petal will allow for the measurement of the second space point co-ordinate.

The R0 module will have two hybrids each reading out two rows of strips. Each row will have an extra strip at each end that will not be readout strips but will rather be used to shape the electric field for the outer readout strips. The two inner rows will have a different angular pitch to the two outer rows, where the pitch is the distance from the center of one strip to the center of the next strip. The thickness of the sensor will be 310μm with an allowed error of up 25μm and more R0 specifics are given in Table 1[7].

3. Testbeam

The testbeam telescope is a EUDET-type telescope[8] operated by the EUDAQ framework and analysed with the EUTelescope software[9]. EUDET was the detector R&D towards the

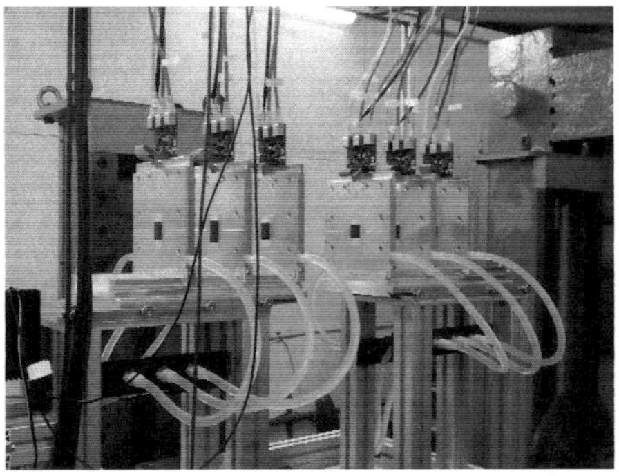

Figure 3. Image of the EUDET testbeam telescope at DESY[9]. Shown are the six mimosa pixel detectors used for beam tracking when a device is being tested, which would be placed in the gap in the middle of the telescope. The beams would come in from the right where the large red object is part of the magnet system which focuses the beam.

International Linear Collider(ILC) which ended in 2010. The telescope is located at DESY in Hamburg, Germany and is where the R0 module will undergo testing.
The telescope comprises of Mimosa26 high granularity pixel detectors[10], as shown in Fig. 3, that are used for track fitting. The sensor that is being tested is known as a DUT (Device Under Test) and is placed between the third and fourth mimosa detectors. The beam tracking can be performed to precisions of up to 2 μm[9], usually much smaller than the DUT resolution, with the telescope utilising electron beams which are produced through the conversion of bremsstrahlung beams originating from carbon fibre targets in the DESY II accelerator. The energy of the electrons range from 1 GeV to 6 GeV[9] and are used to test whether a hit in the DUT corresponds to a track calculated from hits in the mimosa planes passing through the sensor at that point.

4. AllPix

AllPix[11] is a Geant4[12] based simulator dedicated to the study of solid state detectors where it is possible to create an experimental set-up of how ever many detectors one would like and in any configuration, so long as it's within the limits of Geant4. Geant4 is a C++ toolkit used for the simulation of particles through matter based on the particles energy and the radiation length of the material. It is in this software that the geometries are defined, as well as where the physics processes and particles used are chosen. The AllPix software is used to simulate the EUDET telescope and the digitisers that are used to digitise the hits from Geant4. This simulated data obtained from AllPix will then be analysed as if coming from the telescope itself. The individual detector specifications are defined in a separate file using eXtensible Markup Language which is called by a macro file within which the experimental set-up is defined and the simulation is run from.

5. EUTelescope

EUTelescope[13] is a generic pixel telescope data analysis framework comprising of a group of Marlin processors and is embedded in the ILCsoft framework. The software is used to extract tracks from LCIO data (Linear Collider I/O) and it is used in the analysis of data from both the testbeam telescope as well as AllPix simulation. The raw data from either the experiment or simulation is converted into LCIO format using a suitable converter. For the experimental data, a converter in the EUTelescope package is available, whereas a converter that comes as

part of the AllPix package is used for converting the AllPix data.
Clustering, alignment and track fitting is performed on the LCIO data to obtain the tracks of particles through the telescope to compare with a DUT hit. During these steps a GEAR file (GEometry API for Reconstruction) is called within which the geometry and orientation of the detectors is defined. These geometries may be defined using the geometry class TGeo of CERN's data analysis framework, ROOT[14], in a separate file and called by the GEAR file as well. Once the tracks have been extracted from the data, they are analysed to check whether a registered hit in the DUT corresponds to a particle passing through the DUT at that point.

6. Current Status of R0 Module Simulation

The current designs of the simulated geometries for the software packages AllPix and EUTelescope are shown in Fig. 4 while Fig. 5 shows the design within the telescope as a DUT. The EUTelescope design is simpler than the AllPix design as AllPix will actually simulate how the particles interact with the sensor and digitise the hits obtained from that, whereas EUTelescope will perform more of a mapping from the hits in the LCIO data to the geometry defined using the TGeo class. The main characteristic these designs do not share with the R0 module is the stereo annulus shape as these designs are basic segments of a cylinder. The reason for this is that work is still being done on how to implement the stereo annulus shape in these software packages.

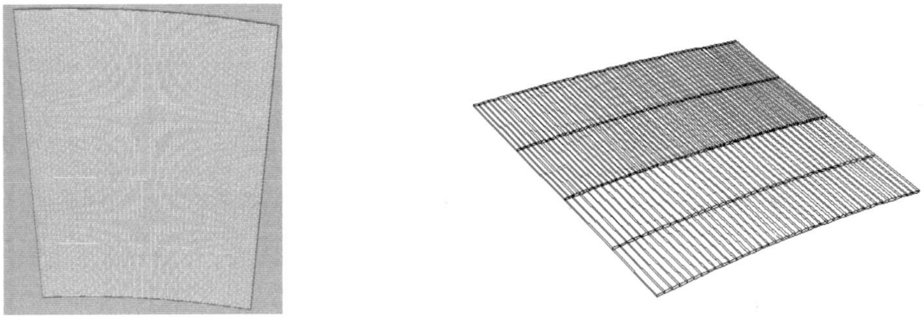

(a) AllPix simulation (b) EUTelescope simulation

Figure 4. Current progress of the simulated R0 module. Both images do not have the correct amount of strips for visualisation purposes as the strips would not be observable if the correct number of strips were used.

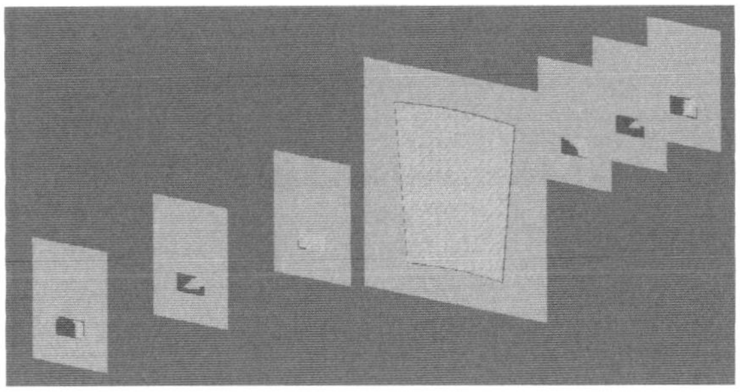

Figure 5. Image of the simulated EUDET telescope with the R0 module placed as a DUT. Placement of the planes along the z-axis is 0, 59, 119.5, 250, 404, 534 and 604 mm.

IOP Conf. Series: Journal of Physics: Conf. Series **889** (2017) 012017 doi:10.1088/1742-6596/889/1/012017

7. Conclusion

Due to the HL-LHC upgrade, the ATLAS detector will require an upgrade to cope with the new high pile-up environment. However before the actual upgrade can commence, research and development has to occur. Part of this R&D is the characterisation of the detector and the sensors that will be used in simulations. Progress on the R0 module of the strip endcap in the ITk so far has provided a radial geometry with specifications of the R0 module. However the stereo annulus shape has not yet been simulated due to its complicated geometry. Once a decision on the final shape has been made, a simulation of the geometry in AllPix is required from which data will be obtained that will be analysed in EUTelescope.

Acknowledgments

I would like to thank my supervisor Dr. Sahal Yacoob as well as my co-supervisors Dr. Kenneth G. Wraight, Dr. Stephen W. Peterson and Dr. Andrew Blue for their important input and assistance, Prof. Bruce Mellado and Prof. Alan Cornell of the University of the Witwatersrand for organising the High Energy Particle Physics workshop held at iThemba labs in Gauteng and finally the NRF for funding my accommodation and transport during my stay at the workshop.

References

[1] ATLAS Collaboration (2012) Observation of a new particle in the search for the Standard Model Higgs boson with the ATLAS detector at the LHC, *Phys. Lett.* **B716** 1-29. Online: [arXiv:1207.7214[hep-ex]]

[2] CMS Collaboration (2012) Observation of a new boson at a mass of 125 GeV with the CMS experiment at the LHC, *Phys. Lett.* **B716** 30-61. Online: [arXiv:1207.7235[hep-ex]]

[3] ATLAS Collaboration (2016) Technical Design Report for the ATLAS ITk Strip Detector, ATL-COM-UPGRADE-2016-040. Online: https://cds.cern.ch/record/2239048.

[4] The ATLAS Collaboration (2012) Letter of Intent for the Phase-II Upgrade for the ATLAS Experiment, (Geneva: *Tech. Rep.* **CERN-LHCC-2012-022; LHCC-I-023**). Online: https://cds.cern.ch/record/1502664.

[5] ATLAS collaboration (2008) The ATLAS Experiment at the CERN Large Hadron Collider, *JINST* **3** S08003. Online: http://iopscience.iop.org/article/10.1088/1748-0221/3/08/S08003/meta

[6] Noemi Calace on behalf of the ATLAS collaboration (2016) ATLAS ITk Layout Design and Optimisation, ATL-UPGRADE-SLIDE-2016-867. Online: https://cds.cern.ch/record/2234847

[7] ATLAS Upgrade Strip Sensor Collaboration (2016) Supply of Silicon Microstrip Sensors of ATLAS12EC specifications. Online: https://edms.cern.ch/document/1547701/2.2

[8] Jansen, H. *et. al.* (2016) Performance of the EUDET-type beam telescopes. Online: [arXiv:1603.09669 [physics.ins-det]].

[9] Dreyling (2015) Welcome to EUDET-type beam telescopes. Online: https://telescopes.desy.de/Main_Page

[10] Winter, M. (2009) Development of Swift, High Resolution, Pixel Sensor Systems for a High Precision Vertex Detector suited to the ILC Running Conditions, DESY PRC R&D Nr 01/04. Online: http://www.iphc.cnrs.fr/IMG/desy_mimosa_report_prc09.pdf

[11] Benoit, M. Idarraga, J. (2014) The AllPix Simulation Framework. Online: https://twiki.cern.ch/twiki/bin/view/Main/AllPix

[12] Geant4 Collaboration (2003) GEANT4: a simulation toolkit, *Nucl. Instrum. Meth.* A **506** 250-303. Online: http://inspirehep.net/record/593382/files/fermilab-pub-03-339.pdf

[13] Bulgheroni, A. Klimkovich, T. Roloff, P. and Zarnecki, A.F. (2007) EUTelescope: tracking software, EUDET Memo-2007-20. Online: https://www.eudet.org/e26/e28/e182/e425/eudet-memo-2007-20.pdf

[14] Brun, R. Rademakers, F. (1997) ROOT - An Object Oriented Data Analysis Framework, (Lausanne: *Nucl. Instrum. Meth.* A **389**) p 81-86. Online: http://www.sciencedirect.com/science/article/pii/S016890029700048X

IOP Conf. Series: Journal of Physics: Conf. Series **889** (2017) 012018 doi:10.1088/1742-6596/889/1/012018

Measurement of open heavy-flavour production as a function of charged-particle multiplicity with ALICE at the LHC

Sibaliso Mhlanga
University of Cape Town, Rondebosch, Cape Town, South Africa
iThemba LABS, Sommerset West, Western Cape
For the ALICE Collaboration

E-mail: sibahso.mhlanga@cern.ch

Abstract. Heavy quarks are produced in the early stages of ultra-relativistic hadron collisions via hard scatterings and are an important tool for studying different aspects of Quantum Chromodynamics (QCD) in hadronic collisions. Charged-particle multiplicity gives information on the global characteristics of the event and could be used to characterize particle production mechanisms. In hadronic collisions at Large Hadron Collider (LHC) energies, there is a significant contribution of multi-parton interactions. The measurement of heavy-flavour yields as a function of charged-particle multiplicity gives insight into the mechanisms influencing their production in hadronic collisions at these energies and it is a tool to test the possible influence of multi-parton interactions. Furthermore, the charged-particle multiplicity dependence of open heavy flavours is used to test the ability of QCD theoretical models to describe the data. In ALICE, heavy-flavour production is measured via the hadronic and semi-leptonic decay channels (electrons at central rapidity and muons at forward rapidity). Charged-particle multiplicity is measured at central and forward rapidity. We will present the results on open heavy-flavour production as a function of the charged-particle multiplicity in pp and p–Pb collisions. Results will be compared to quarkonia measurements as well as theoretical model calculations.

1. Introduction

Heavy quarks (charm and beauty) are abundantly produced at the LHC [1] in the early stages of hadronic collisions via hard scattering processes and experience the full evolution of the collision. The measurement of heavy flavours in pp collisions can be used to test pQCD calculations. In addition, they provide an essential baseline for the studies in nuclear (p–Pb and Pb–Pb) collisions. Their production in nuclear collisions is modified by cold nuclear matter effects (CNM) such as shadowing and energy loss [2, 3]. The knowledge of these effects is fundamental for understanding the interactions of heavy quarks with the deconfined medium formed in heavy-ion collisions where the modified transverse momentum distribution can be used to infer the medium's characteristics. Heavy quark production is experimentally accessible through the measurement of heavy-flavour hadrons (D or B hadrons) via their decay products.

The charged-particle multiplicity is defined as the average number of charged particles produced per event. It is a key observable for multi-parton interactions (MPI) in high-energy hadronic collisions [4, 5]. The charged-particle multiplicity may be used to test models of particle production such as PYTHIA [6] and PHOJET [7]. Finally, the charged-particle multiplicity dependence of open heavy-flavour yields provides insight into mechanisms of particle production in hadronic collisions. In this report we present the relative yield of D mesons as a function of the relative charged-particle multiplicity in pp collisions at $\sqrt{s}$ = 7 TeV [8] and in p–Pb collisions at $\sqrt{s_{NN}}$ = 5.02 TeV [9].

At the LHC, ALICE [10] comprises central detectors, used for the measurement of charged particles as well as photons and a forward muon spectrometer which measures muons. The L3 solenoid magnet provides a field of 0.5 T to the central barrel detectors covering the rapidity $-0.9 < y < 0.9$. The Silicon Pixel Detector (SPD) is used for the determination of the interaction vertex as well as the measurement of charged-particle multiplicity. Particle identification information is given by the Time Of Flight (TOF) and the Time Projection Chamber (TPC). Two arrays of scintillator detectors (V0A and V0C) placed on both sides of the interaction point provide centrality in Pb–Pb collisions as well as trigger information. They are also used for multiplicity determination. The forward muon spectrometer covers the rapidity range $-4 < y < -2.5$. It consists of a composite absorber, a dipole magnet, five tracking stations, a muon filter and two trigger stations. The absorber reduces background muons from the decays of pions and kaons. The dipole magnet provides a horizontal magnetic field perpendicular to the beam axis and is used for charge and momentum determination. The tracking stations are used to determine the trajectories of the muons traversing the detector. The muon filter is mounted at the end of the spectrometer and filters all particles with momentum $p < 4$ GeV/c. The trigger stations are used for muon identification and triggering.

The reconstruction of D mesons is done via their hadronic decay channels: $D^0 \rightarrow K^-\pi^-$(branching ratio, BR = 3.88±0.05 %), $D^+ \rightarrow K^-\pi^+\pi^+$(BR = 9.13±0.19 %), $D^{*+} \rightarrow D^0\pi^+$ (BR = 67.7±0.05 %) followed by $D^0 \rightarrow K^-\pi^+$[11] and their charge conjugates at central rapidity. Further details on D meson and charged-particle multiplicity measurements are reported in [8] and [12]. The measurement of prompt and non-prompt (produced via the decay of beauty hadrons) $J/\psi \rightarrow e^+e^+$ (BR = 5.97±0.03 %) and $J/\psi \rightarrow \mu^+\mu^+$ (BR = 5.96±0.03 %) at central and forward rapidity are reported in [13].

2. Results

2.1. D-meson production in pp collisions

The results of the relative yields (ratio between the yield in a given multiplicity interval normalised to the multiplicity-integrated one) of D mesons, D^0, D^+, D^{*+}, measured at central rapidity as a function of relative charged-particle multiplicity (multiplicity of charged particles normalised to the average value) in pp collisions at $\sqrt{s}$ = 7 TeV are presented in Figure 1. In Figure 1(a), the prompt D-meson relative yield is compared with that of inclusive J/ψ as a function of the relative multiplicity. The yields of D mesons and inclusive J/ψ show a similar increase with the multiplicity, in particular for J/ψ measured at central rapidity. In Figure 1(b) the yield of D mesons is compared to that of non-prompt J/ψ (produced in the decay of beauty hadrons), which provides an indirect measurement of the multiplicity dependence of beauty hadrons. The trend is similar to that observed in Figure 1(a). We observe no difference, within uncertainty, in the multiplicity dependence of charm and beauty hadron yields at central rapidity.

2.2. D-meson production in p–Pb collisions

Figure 2 shows the prompt D-meson yield as a function of the relative multiplicity measured at central rapidity ($-0.9 < y < 0.9$) compared with J/ψ in p–Pb collisions at $\sqrt{s_{NN}}$ = 5.02 TeV

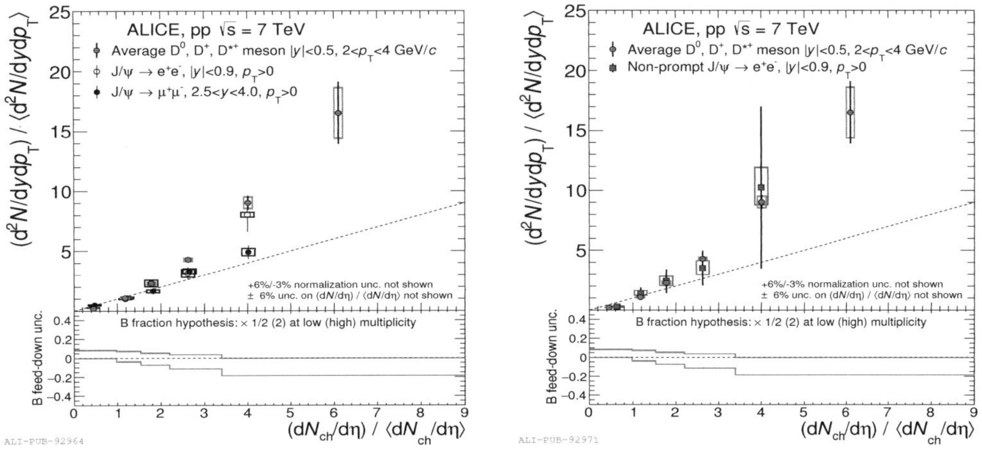

(a) (b)

Figure 1. Average D meson yield as a function of the relative charged-particle multiplicity in pp collisions at $\sqrt{s}$ = 7 TeV, measured at central rapidity in the interval $2 < p_{\mathrm{T}} < 4$ GeV/c compared to the yields of inclusive J/ψ (a) and non-prompt J/ψ (b) for $p_{\mathrm{T}} > 0$ GeV/c [8, 13]. The vertical bars and the boxes represent the statistical and systematic uncertainties respectively. The bottom panels show the B feed-down fraction uncertainty.

at forward (p going) and backward (Pb going) rapidity. The yield of D mesons increases faster than the J/ψ yield towards high multiplicity, particularly for J/ψ measured at forward rapidity.

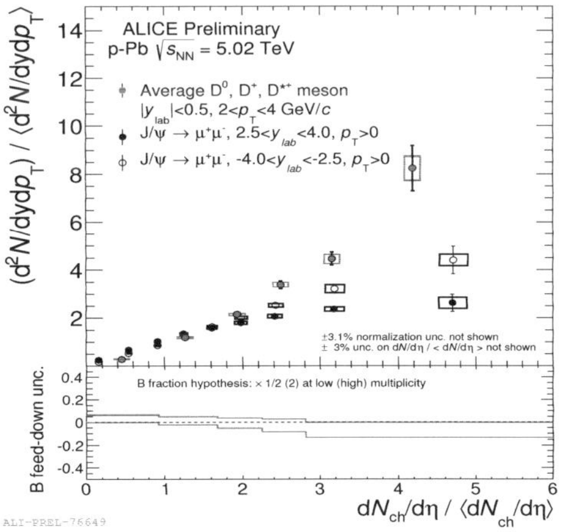

Figure 2. Relative yields of the average of the D-meson species and J/ψ in the forward (p going) and backward (Pb going) rapidity as a function of the relative charged-particle multiplicity [9, 13]. The vertical bars and the boxes represent the statistical and systematic uncertainties respectively. The bottom panel shows the B feed-down fraction uncertainty.

IOP Conf. Series: Journal of Physics: Conf. Series **889** (2017) 012018 doi:10.1088/1742-6596/889/1/012018

2.3. *Comparison of D-meson measurements in pp and p–Pb collisions*

The yield of D mesons in pp collisions at $\sqrt{s} = 7$ TeV is compared to the yield in p–Pb collisions at $\sqrt{s_{NN}} = 5.02$ TeV as a function of charged-particle multiplicity in Figure 3. The D-meson yield increases faster than linear with multiplicity for both pp and p–Pb collisions. The increase is independent of the colliding system.

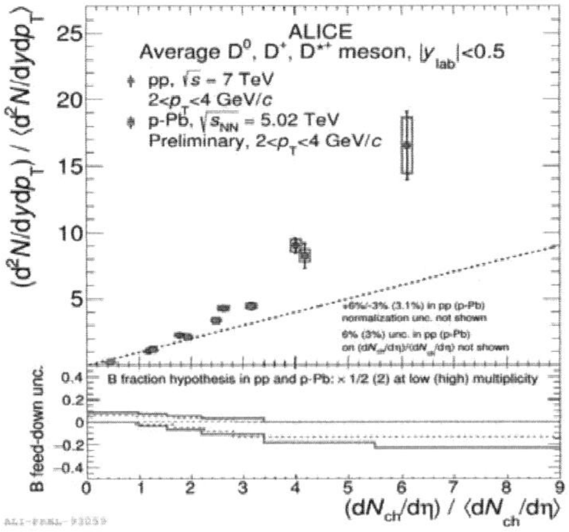

Figure 3. The relative yields of the averages of the three D-meson species as a function of the relative charged-particle multiplicity in pp and p–Pb collisions at $\sqrt{s} = 7$ TeV and $\sqrt{s_{NN}} = 5.02$ TeV, respectively [9]. The vertical bars and the boxes represent the statistical and systematic uncertainties respectively. The bottom panel shows the B feed-down fraction uncertainty.

2.4. *Comparison of results with theoretical models*

The comparison of the measurements in pp and p–Pb collisions with theoretical calculations is shown in Figure 4. In pp collisions the yield of D mesons as a function of the charged-particle multiplicity is compared with theoretical predictions based on percolation [14, 15], EPOS [16] with and without hydrodynamical effects and PYTHIA 8 [17]. All the models predict an increase of the relative yield of D mesons with the relative multiplicity for both pp and p–Pb collisions. However, percolation and EPOS+hydro predict a faster than linear increase while PYTHIA 8 and EPOS are consistent with the data at lower multiplicity. In p–Pb collisions, the results are compared with EPOS+hydro and EPOS. Both models predict an increase of the relative yield with the multiplicity. EPOS+hydro predicts a faster than linear increase.

3. Conclusion

The ALICE Collaboration measured the relative yield of D mesons as a function of the charged-particle multiplicity in pp and p–Pb collisions and compared the results with analogous measurements of inclusive and non-prompt J/ψ. In both colliding systems the relative yield of D mesons increases with multiplicity and the increase is faster than linear. The increase is independent of the $p_{\rm T}$ interval considered. J/ψ yields show a similar increase in pp collisions, however in p–Pb collisions D-meson yields increase faster than J/ψ, particularly at forward rapidity. The results were also compared with theoretical calculations. All models considered in the study predict an increase of heavy-flavour yields with multiplicity. The multiplicity dependence of D-meson production is described better by calculations that include hydrodynamics, especially at higher multiplicity.

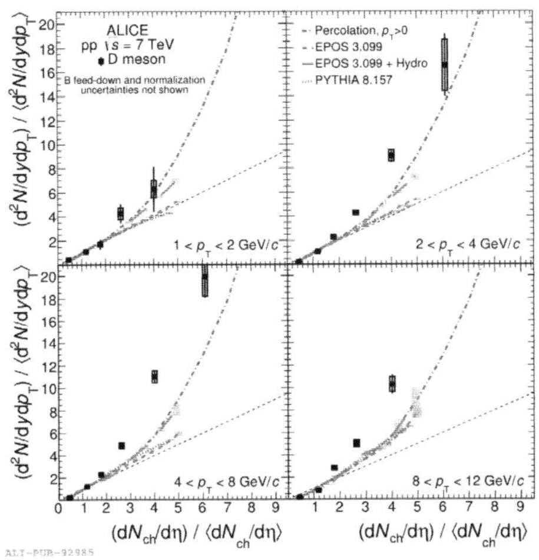

(a)

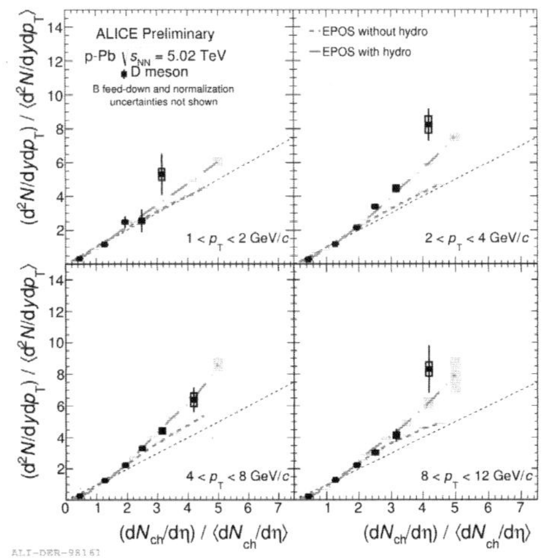

(b)

Figure 4. Relative yields of D mesons in different p_T intervals as a function of the relative charged-particle multiplicity in pp collisions at $\sqrt{s}$= 7 TeV (a) and p–Pb collisions at $\sqrt{s}$ = 5.02 TeV (b). The lines represent various theoretical models [14] - [17].

References

[1] ALICE Collaboration. *Journal of Instrumentation*, 3:S08001, 2008.
[2] S. Gavin and J. Milana. *Physics Review Letters 68 (1992) 1834.*
[3] S. Brodsky and P Hoyer. *Physics Letters B 298 (1993) 165.*
[4] T. Sjöstrand and M. van Zijl. *Physical Review D 36, 2019 (1987).*
[5] P. Bartalini *et al. arXiv:1003.4220 [nucl-ex].*
[6] Sjöstrand Torbjorn. *et al. Journal of High Energy Physics 0605:026,* 2006.
[7] R. Engel and J. Ranft. *Physical Review D 54, 1996.*
[8] ALICE Collaboration. *Journals of High Energy Physics 09 (2015) 148.*
[9] ALICE Collaboration. *arXiv:1602.07240v1 [nucl-ex].*
[10] ALICE Collaboration. *International Journal of Modern Physics A20:1430044,* 2014.
[11] Particle Data Group. *Review of Particle Physics, Chinese Physics C 38 (2014) 090001.*
[12] ALICE Collaboration. *Journals of High Energy Physics 01 (2012) 128.*
[13] ALICE Collaboration. *Physics Letters B 712 (2012) 165-175.*
[14] E. G. Ferreiro and C. Pajares. *Physical Reviews C 86 (2012) 034903.*
[15] E. G. Ferreiro and C. Pajares. *arXiv:1501.03381 [hep-ph].*
[16] H.J. Drescher, M. Hladik, S. Ostapchenko, T. Pierog, K. Werner. *Physics Reports 350 (2001) 93-289.*
[17] K. Werner, B. Guiot, L. Karpenko and T. Pierog. *Physical Reviews C 89 (2014) 064903.*

The influence of neutron radiation damage on the optical properties of plastic scintillator UPS 923A

Skhathisomusa Mthembu[a,1], **Yuri Davydov**[b,2], **Vladimir Baranov**[b,3], **Bruce Mellado Garcia**[a,4], **Joyful Mdhluli**[a,5], **Elias Sideras-Haddad,**[a,6]

[a] School of Physics, University of the Witwatersrand, Johannesburg, Wits 2050, South Africa.
[b] Joint Institute for Nuclear Research, Dubna, Russia, 141980.

E-mail: [1]skhathisomusa.mthembu@cern.ch, [2]davydov@jinr.ru, [3]vladimir.baranov@cern.ch, [4]bruce.mellado.garcia@cern.ch, [5]joyemmie@gmail.com, [6]elias.sideras-haddah@wits.ac.za

Abstract. Plastic scintillators are vital in the reconstruction of hadronic particle energy and tracks resulting from the collision of high energy particles in the Large Hadron Collider (LHC) at CERN. These plastic scintillators are exposed to harsh radiation environments and are susceptible to radiation damage. The effects of radiation damage on the transmittance, luminescence and light yield of Ukraine polystyrene-based scintillator UPS 923A were studied. Samples were irradiated with fast neutrons, of varying energies and fluences, using the IBR-2 reactor FLNP (Frank Laboratory for Nuclear Problems) at the Joint Institute for Nuclear Research. Results show a small change in the transmittance of the higher energy visible spectrum, and a noticeable change in the light yield of the samples as a result of the damage. There is no change observed on the luminescence as a result of radiation damage at studied fluences. The doses and fluences of the neutrons shall be increased and changes in optical properties as a result of the radiation shall be further studied.

1. Introduction

The very nature of detection depends on the mechanisms governing the interactions between incident radiation and the detector material. Scintillators are materials that exhibit luminescence under the presence of ionising radiation. As a result of this luminescence, radiation measurements could be performed, historically, by observing the brightness and intensity of the flashes produced by the radiation incident on the scintillator [1]. There are two common types of scintillators, namely organic and inorganic, with primary differences being structural and from their interactions with ionising radiation. Inorganic scintillators are crystals made from oxides such as bismuth germanate (BGO), or alkali halides such as sodium iodide (NaI) and cesium iodine (CsI). Organic scintillators, or plastic scintillators, as elaborated on in this paper, are of particular interest because of their use in high energy physics for the indirect measurement of the momentum and charge of particles incident on and traversing through them. The plastic scintillator UPS 923A is investigated, because of it's use in the CDF detector at the Fermilab Tevatron [2]. It consists of a polystyrene base/bulk material doped with 2% p-terphenyl (PTP) as the primary fluor and 5-phenyl-2-[4-(5-phenyl-1,3- oxazol-2-yl)phenyl]-1,3-oxazole (POPOP) as a secondary fluor that acts as a wavelength shifter for the sensitivity of the photomultiplier tubes (PMTs) collecting the scintillation light [3]. Large scintillator and

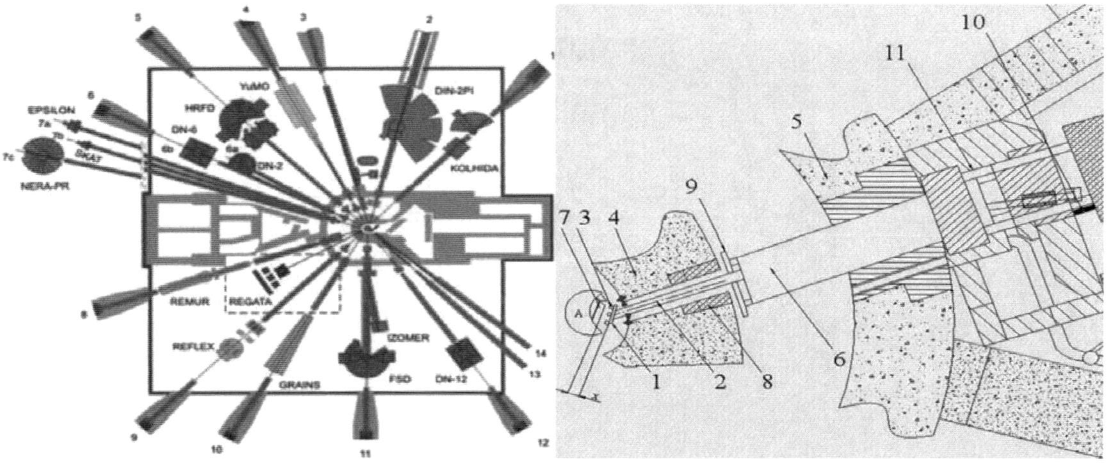

(a) (b)

Figure 1: Schematic diagram of IBR-2 reactor at JINR for neutron irradiation of the samples (a) and a schemtatic of the irradiation facility (b): 1 is the metallic container to hold the samples; 2 the transport beam; 3 are samples for irradiation; 4 is the first biological shield; 5 the second biological shield; 6 is the massive part of irradiation facility; 7 the water moderator; A is the active zone of the IBR-2 reactor; x the distance of moderator surface to the samples, 8 is shutter of biological shield; 9 the emergency stops; 10 is a seen mechanical stop, 11 the rail way. All hatched elements on the figure are parts of the biological shield off the irradiation.

other electronic arrangements [4] relay these measurements' information to complex detection systems for the reconstruction of these particles' trajectories, momenta, and other parametric information of interest, as done by the Tile Calorimeter(TileCal) of ATLAS (A Toroidal LHC Apparatus) at CERN (Geneva, Switzerland) [5].
The optical properties of interest are the transmittance, luminescence and light yield of the scintillators. The transmission study observes how much light, and at what wavelengths, is absorbed as a result of radiation dosage, the luminescence looks at how the scintillating ability (or flourescence) of the material is affected by the varying radiation fluences at varying wavelengths, and the light yield is a more involved study of how the flourescence of a sample is affected by radiation damage under ionising radiation from a specific source [6].

2. Method Overview

The optical properties of Ukraine polystyrene-based plastic scintillator (UPS 923A) were tested by irradiating three of four identical samples of dimensions 20mm · 20mm · 6mm with fast neutrons of fluences $1.8 \cdot 10^{14}$, $1.7 \cdot 10^{13}$ and $3.8 \cdot 10^{12}$ neutrons /cm^2, and studying the change in transmittance, luminescence and light yield of the samples as a function of the radiation dose. The irradiation of samples was performed on channel no. 3 of the IBR-2 reactor at the JINR (Joint Institute for Nuclear Research), Dubna, Russia, as shown in Figure 1 [7]. Three of the four samples were placed at increasing distances from the neutron source, for irradiating at fluences stated above.

The transmittance experiment was performed at the Joint Institute for Nuclear Research (JINR). Samples were placed in a `SHIMADZU SolidSpec-2700DUV` spectrophotometer[8]. Light from 300nm to 800nm, at 1nm intervals, was transmitted through each sample, and the relative intensity spectra of the transmitted light were observed for analysis.

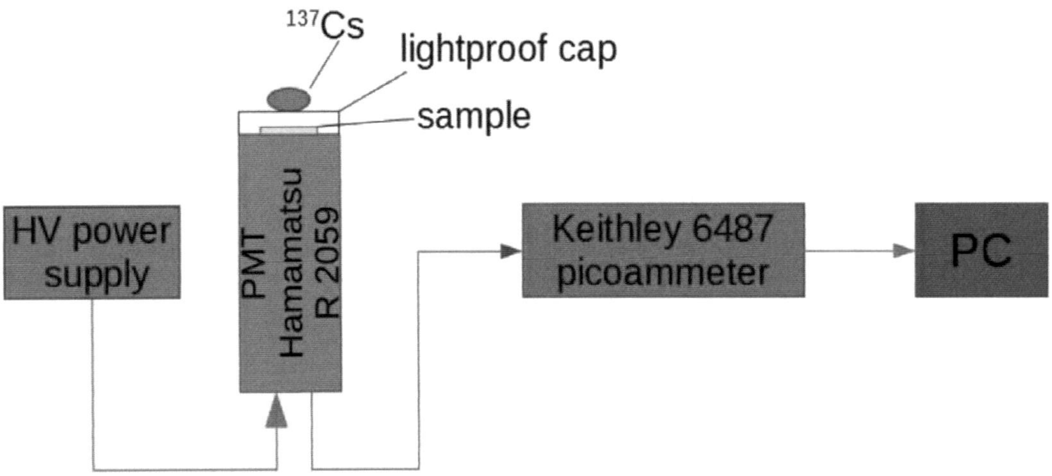

Figure 2: Schematic diagram of the setup for the light yield experiment

The luminescence experiment was conducted at the University of the Wiwatersrand. A laser was shone onto the sample at the Raman Spectrometry Laboratry. A `Lexel 95 SHG` model argon ion laser that uses a frequency-doubling crystal to generate 244nm from the 488nm laser line was used for the luminescence studies. The incident UV laser beam was focused onto the sample using a `Thorlabs` 20x UVB objective. The beam was scanned over a square of size 20micron x 20micron to give some local averaging of the luminescence. The backscattered light was dispersed via a 150 lines/mm grating in a `Horiba LabRAM` spectrograph onto a liquid nitrogen cooled CCD for detection. The accumulation time was 1 second/spectrum and acquired via `LabSpec v5` software. The relative intensity spectra of the luminescence were observed for analysis.
The light yield experiment was performed at the JINR. The experiment was set up as shown on Figure 2

The sample was placed under a light-proof cap with a ^{137}Cs source in close proximity as a gamma source and a `Hamamatsu R 2059` PMT for the detection of the sample's fluorescence. The signal from the PMT was relayed to a `Keithley 6487` picoammeter and the ampere spectrum from the picoammeter was interpreted by computer software. The picoammeter relative intensity (at given currents) spectra were observed for analysis.

3. Analysis and preliminary results

Transmittance
In this study, the transmittance (%) is defined as the intensity of light collected after transmission through the sample divided by the intensity of the light at that wavelength without the sample (transmittance through air). Figure 3 shows the spectral intensities of transmitted light, at given wavelengths, for each sample. Transmission loss is observed at the low energy UV range of the spectrum, from 400nm to 550nm, for sample 1.

Luminescence
Luminescence, for this experiment, is defined as the intensity of light at a given wavelength arising from the excitement of the sample with a 244nm laser. In Figure 4, a translation increase on the intensity axis is observed for increasing radiation dose, with sample 1 of highest received

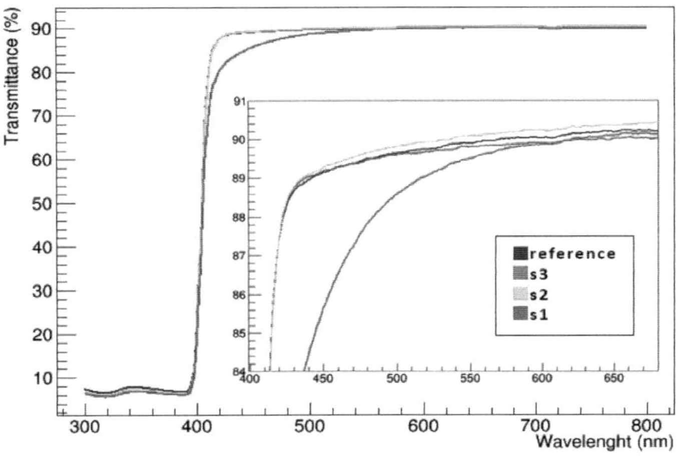

Figure 3: Transmitted intensity per wavelength plot for the four samples where reference is the unirradiated sample (or s0), s3 (sample 3) irradiated at $3.8 \cdot 10^{12}$ neutrons /cm^2, s2 (sample 2) at $1.7 \cdot 10^{13}$ neutrons /cm^2 and s1 (sample 1) at $1.8 \cdot 10^{14}$ neutrons /cm^2.

fluence having the largest intensity and the reference sample having the lowest intensity.

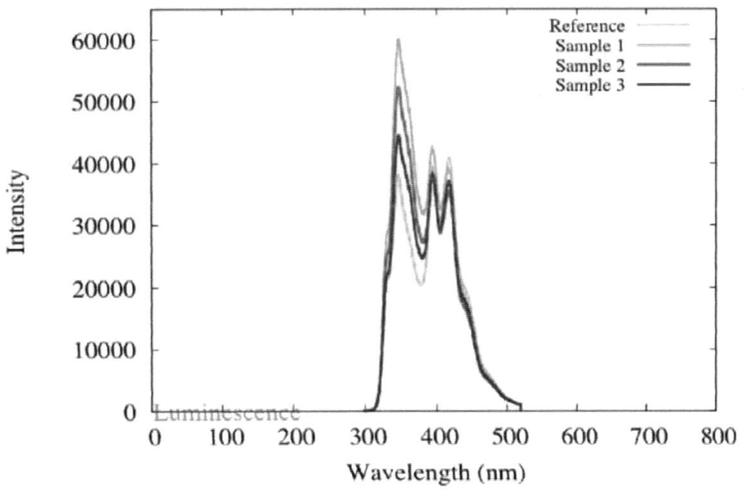

Figure 4: Luminescence per wavelength plot for the four samples where Reference is the unirradiated sample 0, Sample 3 irradiated at $3.8 \cdot 10^{12}$ neutrons /cm^2, Sample 2 at $1.7 \cdot 10^{13}$ neutrons /cm^2 and Sample 1 at $1.8 \cdot 10^{14}$ neutrons /cm^2.

Light Yield

The light yield is defined as the current intensity per unit time, observed for each sample. Where luminescence observes relative energy changes by wavelength, the light yield experiment does so by current output as recorded by the picoammeter and transferred to PC data acquisition software. The data from the experiment are saved as histograms of event numbers per current,

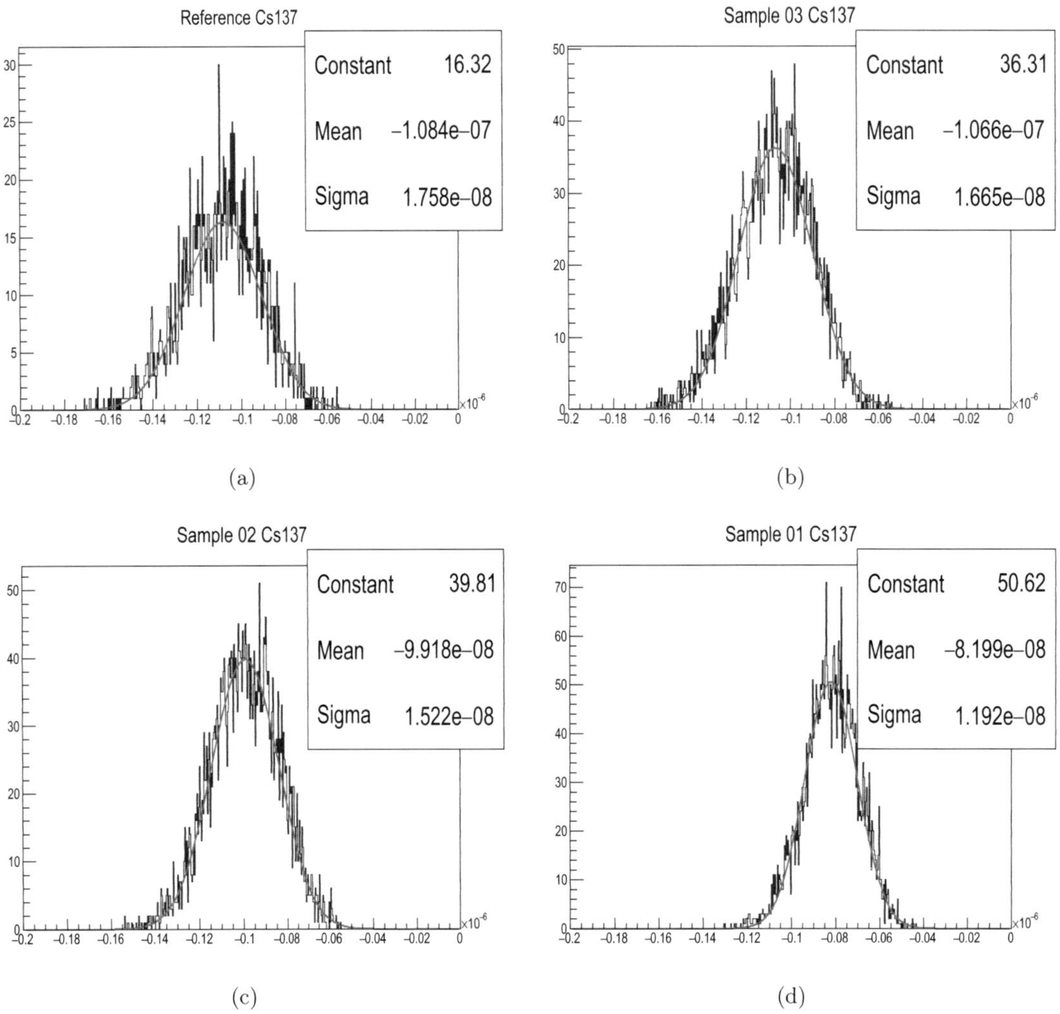

Figure 5: Light yield per wavelength histograms for the four samples where Reference is the unirradiated sample (a), sample 3 irradiated at $3.8 \cdot 10^{12}$ neutrons /cm^2 (b), sample 2 at $1.7 \cdot 10^{13}$ neutrons /cm^2 (c) and sample 1 at $1.8 \cdot 10^{14}$ neutrons /cm^2 (d). The x-axes represent the current, and the y-axes represent the intensity.

and ROOT is used to fit a Gaussian distribution on the histograms to calculate the average current, standard deviation, etc. as shown in Figure 5. The current is correlated to the mean of the Gaussian, and the currents are compared for the different samples.

The mean currents from the different samples were normalised to the reference sample, and the results observed on Figure 6.

4. Summary

The optical properties of irradiated UPS 923A were studied as a function of irradiation damage. The study showed that, at the fluences achieved for the irradiation of the scintillator, a noticeable change in the transmittance is observed at fluences of order $1.8 \cdot 10^{14}$ neutrons /cm^2. The luminescence studies showed no observable radiation damage effects for the doses achieved, and

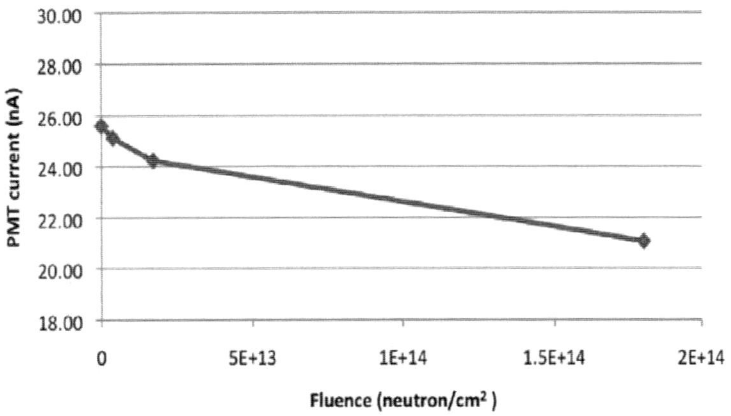

Figure 6: The light yield against irradiation neutron fluence.

the translation of the intensities attributed to surface damages. For a more conclusive study on the effects of radiation damage, higher fluences should be reached. From the light yield studies, the mean current decreased with increasing radiation. Higher radiation fluences should be achieved to establish a direct functional correlation between dosage and radiation damage from neutron radiation. Increasing the dosage will also result in a wider study of radiation damage on the plastic's atomic composition, enriching our appreciation and understanding of plastic scintillators and detection as a whole.

Acknowledgements

We are grateful to S. Kulikov, M. Bulavin and their colleagues from the JINR for sample irradiation help. We acknowledge High Energy Physics Group at the University of the Witwatersrand for fruitful discussions on the scope and direction of the research. S. Mthembu acknowledges funding support from National Research Foundation, South Africa for financial support towards the research work, together with the financial support of the Joint Institute for Nuclear Research.

References

[1] Vasil'chenko, V. G. *et al.*, "New results on radiation damage studies of plastic scintillators," *Nucl. Instr. and Meth. in Phys. Res.* A **369** no. 1, (1996) 55-61.
[2] Artikov, A. *et al.*, "Properties of the Ukraine polystyrene-based plastic scintillator UPS 923A" *Nucl. Instr. and Meth. in Phys. Res.* A **555**, (2005) 125-131.
[3] Wieczorek, A. *et al.*, "A pilot study of the novel J-PET plastic scintillator with 2-(4-styrylphenyl)benzoxazole as a wavelength shifter," `arXiv:1506.00612 [hep-ph]`.
[4] The ATLAS Tile Calorimeter community, "The optical instrumentation of the ATLAS Tile Calorimeter," *J. Instrum.* **8**, no. 1, (2013) P01005.
[5] Jivan, H. *et al.*, *Radiation hardness of plastic scintillators for the Tile Calorimeter of the ATLAS detector.* No. ATL-TILECAL-SLIDE-2014-826.
[6] D'Ambrosio, C. D. *et al.*, *A short Overview on Scintillators.* No. 3a.
[7] Bulavin, M. *et al.*, "Irradiation facility at the IBR-2 reactor for investigation of material radiation hardness" *Nucl. Instr. and Meth. in Phys. Res.* B **343**, (2015) 26-29.
[8] Shimadzu, "UV-VIS-NIR Spectrophotometer," *User Manual* C101-E101D.

IOP Conf. Series: Journal of Physics: Conf. Series **889** (2017) 012020 doi:10.1088/1742-6596/889/1/012020

The Madala hypothesis with Run 1 and 2 data at the LHC

Stefan von Buddenbrock,[a] Alan S. Cornell,[b] Mukesh Kumar[b] and Bruce Mellado[a]

[a]School of Physics, University of the Witwatersrand, Wits 2050, South Africa
[b]National Institute for Theoretical Physics; School of Physics and Mandelstam Institute for Theoretical Physics, University of the Witwatersrand, Johannesburg, Wits 2050, South Africa

E-mail: stef.von.b@cern.ch

Abstract. The Madala hypothesis postulates a new heavy scalar, H, which explains several independent anomalous features seen in ATLAS and CMS data simultaneously. It has already been discussed and constrained in the literature by Run 1 results, and its underlying theory has been explored under the interpretation of a two Higgs doublet model coupled with a scalar singlet, S. When applying the hypothesis to Run 2 results, it can be shown that the constraints from the data are compatible with those obtained using Run 1 results.

1. The Madala hypothesis

Searches for physics beyond the Standard Model (BSM) have become ubiquitous since the discovery of the Standard Model (SM) Higgs boson, h. The discovery of the Higgs boson was the first step in understanding the nature of electroweak symmetry breaking (EWSB). There are currently a plethora of BSM physics scenarios extending the notion of EWSB in the literature, many of these being within the reach of the Large Hadron Collider (LHC) running at an energy of $\sqrt{s} = 13$ TeV.

One such scenario is the *Madala hypothesis*. The Madala hypothesis was formulated in 2015 with the aim of connecting and explaining several anomalies in the data from Run 1 of the LHC [1]. At first, this was done through the introduction of a heavy boson H – the *Madala boson* – with a mass in the range $2m_h < m_H < 2m_t$. It was shown that if the SM Higgs boson could be produced via the decay of H, it would contribute to the apparent distortion of the Higgs p_T spectrum from the Run 1 ATLAS differential distributions [2, 3]. This is achieved through the effective decay vertex of $H \to h\chi\chi$ shown in Figure 1(a), where χ is a scalar dark matter (DM) candidate of mass $\sim \frac{1}{2}m_h$. A fit was performed using a complete set of ATLAS and CMS data (at the time), and the parameters of the model were constrained. The two parameters of interest which were are constrained were the mass of H, to a value of $m_H = 272^{+12}_{-9}$ GeV, and the scaling factor $\beta_g = 1.5 \pm 0.6$ which modifies the gluon fusion (ggF) production cross section of H.

The hypothesis was then extended by introducing a DM mediator S [4], such that the effective decay vertex in Figure 1(a) is resolved into the cascading decay process depicted in Figure 1(b). The mass of S lies in the range $m_h < m_S < m_H - m_h$ in order for the decay process $H \to Sh$ to

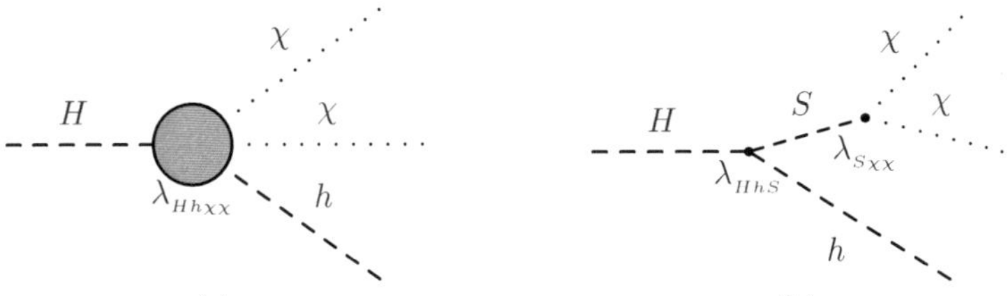

Figure 1: Higgs production in association with DM through the decay of H using (a) an effective vertex and (b) the decay of a DM mediator S.

be on-shell.[1] The H then preferentially decays to one of three pairs of bosons: $H \to SS, Sh, hh$. The S boson was also considered to have small couplings to SM particles, and the assumption that was made is that it can be Higgs-like such that all of its branching ratios (BRs) are already defined. The S would then decay predominantly to the vector bosons Z and $W^{\pm}$ if it has a mass of around 160 GeV or above.

Since these studies were done, however, several new results have been released by the ATLAS and CMS collaborations. In particular, the first Run 2 results for resonant di-Higgs production and di-boson production have been released. In addition to this, both ATLAS and CMS have published Run 1 differential distributions (including Higgs p_{T}) for the $h \to WW \to e\nu\mu\nu$ channel [5, 6], and ATLAS have also released their Run 2 result for $h \to \gamma\gamma$ [7]. It is therefore instructive to determine whether or not these new results are compatible with the Run 1 fit result mentioned above.

2. Exploring resonant search channels

The first check for compatibility is to explore the resonant search channels which contributed to the Run 1 fit result mentioned above. These include resonant searches for $H \to hh$ and VV (where $V = Z, W^{\pm}$).

Since no combination of these results exists, extracting meaningful information from them becomes a matter of statistics. In this study, results are combined through the addition of units of χ^2. For measurements, χ^2 is calculated as Pearson's test statistic. That is, a measurement μ^{exp} along with its associated uncertainty $\Delta\mu^{\mathrm{exp}}$ are tested against a theoretical prediction μ^{th} and its uncertainty $\Delta\mu^{\mathrm{th}}$ by calculating the following:[2]

$$\chi^2 = \frac{(\mu^{\mathrm{th}} - \mu^{\mathrm{exp}})^2}{(\Delta\mu^{\mathrm{th}})^2 + (\Delta\mu^{\mathrm{exp}})^2}. \tag{1}$$

Most results in the literature, however, come in the form of 95% CL limits. In this case, a modified version of Pearson's test statistic is used. Namely, for an observed and expected limit L^{obs} and L^{exp}, respectively, a theoretical prediction μ^{th} can be tested using the following:

$$\chi^2 = \frac{(L^{\mathrm{obs}} - L^{\mathrm{exp}} - \mu^{\mathrm{th}})^2}{(L^{\mathrm{exp}}/1.96)^2}, \tag{2}$$

[1] The S boson is not actually required to decay on-shell from H; this is merely a convenient mass range to simplify phenomenology. This facet has earned it the nickname *the Shelly boson*.

[2] The denominator here differs from Pearson's test statistic, since it already assumes that the theoretical and experimental uncertainties are independent and can therefore be added in quadrature.

Result type	Collaboration	Run	Final state	Luminosity [fb^{-1}]
Higgs p_T spectrum	ATLAS	1	$\gamma\gamma$, $ZZ^* \to 4\ell$, $WW^* \to e\nu\mu\nu$	20.3
	CMS	1	$\gamma\gamma$, $ZZ^* \to 4\ell$, $WW^* \to e\nu\mu\nu$	19.4-19.7
	ATLAS	2	$\gamma\gamma$	13.3
Di-Higgs	ATLAS	1	$bb\tau\tau$, $\gamma\gamma WW^*$, $\gamma\gamma bb$, $bbbb$	20.3
	CMS	1	$bb\tau\tau$, $\gamma\gamma bb$, multilepton	19.5-19.7
	ATLAS	2	$\gamma\gamma bb$	3.2
			$bbbb$, $\gamma\gamma WW$	13.3
	CMS	2	$bbbb$, $bbWW$	2.3
			$\gamma\gamma bb$	2.7
			$bb\tau\tau$	12.9
Di-boson	ATLAS	1	WW, ZZ	20.3
	CMS	1	WW, ZZ	19.7
	ATLAS	2	$ZZ \to 4\ell$	14.8
			$ZZ \to 2\ell 2\nu$	13.3
			$WW \to e\nu\mu\nu$	13.2
	CMS	2	$ZZ \to 2\ell 2\nu$, $WW \to 2\ell 2\nu$	3.2

Table 1: A list of results that are relevant for consideration in the Madala hypothesis. Since results are constantly being updated in newer publications from the experimental collaborations, the integrated luminosity that has been used in this paper has also been tabulated.

where the factor of 1.96 in the denominator arises from the fact that a 95% CL corresponds to 1.96 units of standard deviation.

To understand whether or not a potential signal already lies in the LHC data, the limits coming from the di-Higgs and di-boson searches listed in Table 1 were scanned and evaluated using Equation 2. A best fit value for a production cross section times BR of H was determined as a function of m_H by minimising the sum of χ^2 coming from each search channel. These best fit values as well as 1σ error bands are shown in Figure 2. Note here that since results are shown both at 8 TeV and 13 TeV, the cross sections should not be directly compared between the two energies. However, the reader should keep in mind that the scaling factor from 8 TeV to 13 TeV for the production cross section of H lies between 2.7 and 3.0 as m_H increases from 250 GeV to 350 GeV (calculated using the NNLO+NNLL cross sections in reference [10]). The plots indicate that the best fit value tends to deviate from the null hypothesis (i.e. that H is not produced at all) mostly in the range between 260 GeV and 300 GeV. The only clear exception is in the Run 2 $H \to ZZ$ search results, where the best fit line underestimates even the null hypothesis. However, the $H \to ZZ$ BR was found to be small in the Run 1 fit result [1], and the value presented here is compatible with a small BR within the large uncertainties that surround the central value. This is compatible with the best fit mass of 272 GeV which was obtained in the Run 1 fit result mentioned in section 1. The Madala boson of course can be as light as 250 GeV, but since di-Higgs and di-boson results seldom consider masses below 260 GeV, the scans in Figure 2 are limited.

3. Fitting the Higgs p_T spectrum

Another key aspect of the Run 1 fit result mentioned in section 1 is the Madala hypothesis's ability to predict a distorted Higgs p_T spectrum. In the Run 1 data, this effect is most notably seen in the ATLAS results, where differential distributions are presented in fiducial volumes of phase space [2, 3, 5]. Through the effective decay of H to $h\chi\chi$, the BSM component of Higgs p_T can be added to a SM prediction to reproduce the systematic enhancement of fiducial cross section in the range 20 GeV $< p_T <$ 100 GeV, therefore improving the theoretical description

IOP Conf. Series: Journal of Physics: Conf. Series **889** (2017) 012020 doi:10.1088/1742-6596/889/1/012020

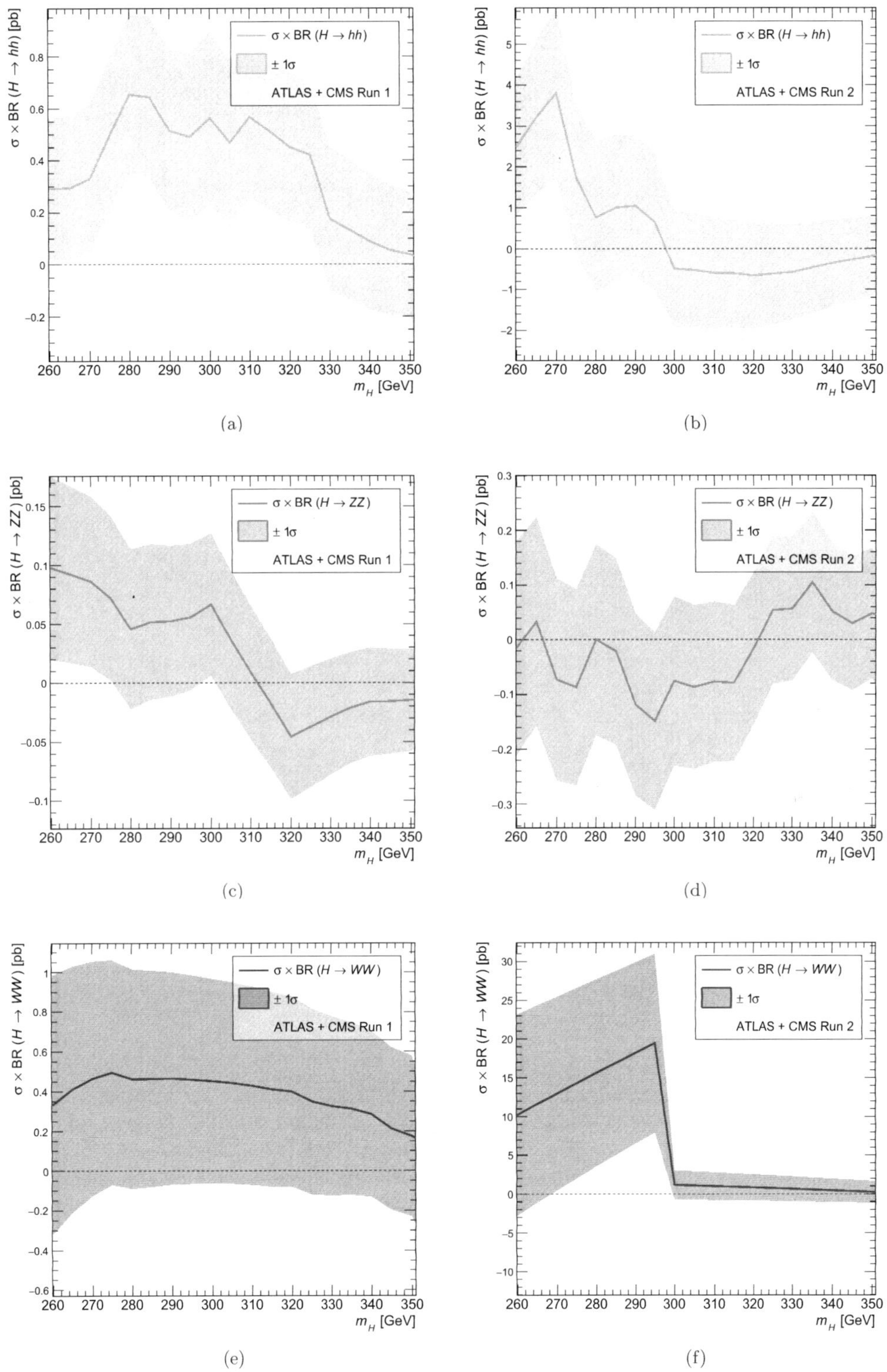

Figure 2: The best fit values of cross section times BR for H production and decay into di-Higgs (top), ZZ (middle), and WW (bottom). The values have been separated into the 8 TeV Run 1 results (left) and the 13 TeV Run 2 results (right). The results on the left and right only include the data exclusively from Run 1 and Run 2, respectively, since the cross sections can not be directly compared at different energies.

p_T spectrum	Data period	Luminosity [fb^{-1}]	β_g
ATLAS $h \to WW$	Run 1	20.3	1.4 ± 0.6
ATLAS $h \to \gamma\gamma$	Run 2	13.3	1.0 ± 0.9
CMS $h \to WW$	Run 1	19.4	0

Table 2: Fit results for the study done on the Higgs p_T spectrum. Here the effective vertex shown in Figure 1(a) was used, with $m_H = 270$ GeV and $m_\chi = 60$ GeV.

of the data.

In order to test whether or not such an improvement can be seen in the results released since the Run 1 fit result, a set of Monte Carlo (MC) samples were made to reproduce the different components of the Higgs p_T spectrum. The SM Higgs p_T spectrum was separated into its different production mechanisms. The ggF spectrum was generated using the NNLOPS procedure [8], which is accurate to next-to-next-to leading order (NNLO) in QCD. The associated production modes – vector boson fusion (VBF), Vh and tth which are commonly labelled together as Xh – were generated at next to leading order (NLO) using MG5_AMC@NLO [9]. These spectra are scaled to the cross sections provided by the LHC Higgs Cross Section Working Group (LHCHXSWG) [10] (from which the theoretical uncertainty also comes). The events are also passed through an even selection identical to the fiducial selection recommended by the experimental collaborations. A further scaling factor was applied to the SM ggF prediction, this being the reported signal strength of ggF (often denoted as $\mu_{gg\text{F}}$).

The BSM prediction (i.e. the Madala hypothesis prediction of $gg \to H \to h\chi\chi$ as shown in Figure 1(a)) was generated using PYTHIA 8.2 [11]. These events were scaled to the LHCHXSWG N^{3}LO ggF cross sections for a high mass Higgs-like scalar, and passed through the fiducial selections as well. Since the Run 1 fit result had a best fit mass of $m_H = 272$ GeV with $m_\chi = 60$ GeV, the mass points considered for this study were $m_H = 270$ GeV and $m_\chi = 60$ GeV.

With the MC samples scaled accordingly each spectrum was added, and a χ^2 value was calculated for each bin per channel, as in Equation 1. The BSM component was scaled such

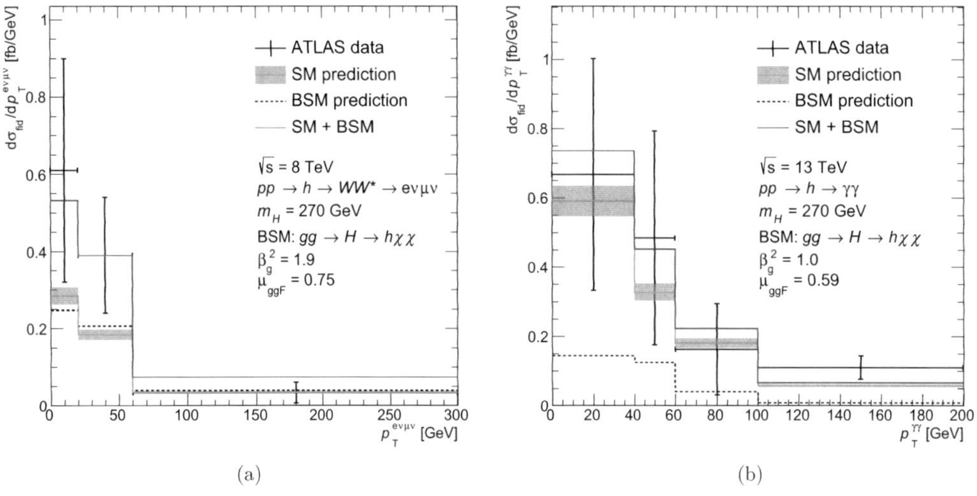

Figure 3: The Higgs p_T spectra for (a) the ATLAS Run 1 $h \to WW$ channel and (b) the ATLAS Run 2 $h \to \gamma\gamma$ channel. The mass points considered here are $m_H = 270$ GeV and $m_\chi = 60$ GeV, as described in the text.

that the total χ^2 was minimised. This BSM scaling is interpreted to be equal to β_g^2, which is a dimensionless factor that multiplies the effective g-g-H coupling, and therefore controls the production cross section of H through ggF. The results of this fit are shown in Table 2. The Run 1 fit result mentioned in section 1 has a value of $\beta_g = 1.5 \pm 0.6$, and here it can be seen that the ATLAS Run 1 $h \to WW$ and ATLAS Run 2 $h \to \gamma\gamma$ results are compatible with this value. The CMS Run 1 $h \to WW$ is not improved by the BSM hypothesis. The p_{T} spectra for the best fit values are shown in Figure 3 for the two spectra which are improved by the BSM hypothesis.

4. Conclusions

The Madala hypothesis was proposed in 2015 as an explanation of several anomalies in experimental data from the LHC. However, since then many newer results have come out which should corroborate the hypothesis if it exists in nature. In this work, these newer results have been tested using a statistical method, and are shown to be compatible with the result obtained in 2015. That is, most of the excesses from Run 1 which motivated the Madala hypothesis have reappeared in the current ensemble of Run 2 results.

However, this ensemble of Run 2 results comprises of preliminary studies which do not make use of the full integrated luminosity which has been accrued by the detectors over the duration of Run 2 at the LHC. It is therefore imperative that a far more detailed study be done when such results become available, since the time is near when enough data will be available to make more definite statements about the Madala hypothesis. Some of the results in this paper are made from experimental plots containing even less than 5 fb^{-1} of data. One would expect to be able to make a statement with $\mathcal{O} \sim 3\sigma$ confidence for individual search channels with at least 50 fb^{-1} of data. Until such a time arrives, the phenomenology Madala hypothesis shall continue to be studied in the context of various BSM scenarios, to gain an understanding of how we might treat it in future.

References

[1] von Buddenbrock S, Chakrabarty N, Cornell A S, Kar D, Kumar M, Mandal T, Mellado B, Mukhopadhyaya B and Reed R G 2015 (`arXiv: 1506.00612`)
[2] Aad G *et al.* (ATLAS Collaboration) 2014 *JHEP* **09** 112 (`arXiv: 1407.4222`)
[3] Aad G *et al.* (ATLAS Collaboration) 2014 *Phys. Lett.* **B738** 234–253 (`arXiv: 1408.3226`)
[4] von Buddenbrock S, Chakrabarty N, Cornell A S, Kar D, Kumar M, Mandal T, Mellado B, Mukhopadhyaya B, Reed R G and Ruan X 2016 *Eur. Phys. J.* **C76** 580 (`arXiv: 1606.01674`)
[5] Aad G *et al.* (ATLAS Collaboration) 2016 *JHEP* **08** 104 (`arXiv: 1604.02997`)
[6] Khachatryan V *et al.* (CMS Collaboration) 2016 *Submitted to: JHEP* (`arXiv: 1606.01522`)
[7] The ATLAS collaboration 2016 ATLAS-CONF-2016-067
[8] Hamilton K, Nason P and Zanderighi G 2015 *JHEP* **05** 140 (`arXiv: 1501.04637`)
[9] Alwall J, Herquet M, Maltoni F, Mattelaer O and Stelzer T 2011 *JHEP* **06** 128 (`arXiv: 1106.0522`)
[10] de Florian D *et al.* (LHC Higgs Cross Section Working Group Collaboration) 2016 (`arXiv: 1610.07922`)
[11] Sjostrand T, Ask S, Christiansen J R, Corke R, Desai N, Ilten P, Mrenna S, Prestel S, Rasmussen C O and Skands P Z 2015 *Comput. Phys. Commun.* **191** 159–177 (`arXiv: 1410.3012`)
[12] Khachatryan V *et al.* (CMS Collaboration) 2014 *JHEP* **09** 087 [Erratum: JHEP10,106(2014)] (`arXiv: 1408.1682`)
[13] Aad G *et al.* (ATLAS Collaboration) 2015 *Phys. Lett.* **B749** 519–541 (`arXiv: 1506.05988`)
[14] Khachatryan V *et al.* (CMS Collaboration) 2016 CMS-PAS-HIG-16-022
[15] Aad G *et al.* (ATLAS Collaboration) 2016 ATLAS-CONF-2016-058

IOP Conf. Series: Journal of Physics: Conf. Series **889** (2017) 012021 doi:10.1088/1742-6596/889/1/012021

Computing the $qg \to qg$ cross section using the BCFW recursion and introduction to jet tomography in heavy ion collisions via MHV techniques

Tanjona R Rabemananajara[1], **W A Horowitz**[2]
[1]African Institute for Mathematical Sciences, 6-7 Melrose Road, Muizenberg, 7945 Cape Town, South Africa.
[2]Department of Physics, University of Cape Town, Private Bag X3, Rondebosch 7701, Cape Town, South Africa.

E-mail: radonirina@aims.ac.za, wa.horowitz@uct.ac.za

February 2017

Abstract. To make predictions for the particle physics processes, one has to compute the cross section of the specific process as this is what one can measure in a modern collider experiment such as the Large Hadron Collider (LHC) at CERN. Theoretically, it has been proven to be extremely difficult to compute scattering amplitudes using conventional methods of Feynman. Calculations with Feynman diagrams are realizations of a perturbative expansion and when doing calculations one has to set up all topologically different diagrams, for a given process up to a given order of coupling in the theory. This quickly makes the calculation of scattering amplitudes a hot mess. Fortunately, one can simplify calculations by considering the helicity amplitude for the *Maximally Helicity Violating* (MHV). This can be extended to the formalism of on-shell recursion, which is able to derive, in a much simpler way the expression of a high order scattering amplitude from lower orders.

1. Introduction

Shortly after the Big Bang, for a few millionths of a second, the universe was filled with an ultra-hot, dense soup (Plasma) of elementary particles mostly composed of quarks and gluons moving at near the speed of light. These conditions have been recreated at powerful accelerators by making head-on collisions between massive ions. The first evidence for jets was seen in 2003 at Brookhaven National Laboratory's Relativistic Heavy Ion Collider (RICH) in the United States and recent experiments at CERN's Large Hadron Collider (LHC) have confirmed the phenomenon of jet quenching in heavy-ion collisions. The study of the radiative energy loss in Quark-Gluon Plasma (QGP) provides insight into the dynamics of the early universe. In heavy ion collisions, the interaction between a high energy quark and a thermal gluon can yield the emission of gluon radiations and the decrease of the energy of the quark [1]. The aim of the project is to compute the cross section for $qg \to qg + ng$ process with $n = 0, 1$ (and more, time permitting). Due to the $n!$ increase in complexity of the usual Feynman approach, novel mathematical structures in gauge theory scattering amplitudes will be used such as Maximally Helicity Violating (MHV) techniques and Britto-Feng-Cachazo-Witten (BCFW) on-shell recursion relations. Indeed, in quantum field theory, one adds up all the Feynman diagrams while each diagram represents a possible collision with an associated probability, and the redundancy due to the gauge invariance leads to a complicated calculation [2]. Recently, significant simplifications in the calculation of scattering amplitudes have been achieved by using new representations for vector polarizations, and the recursion relations. When doing the computation of the $qg \to qg$ scattering amplitude , we will have a look at the case where the conservation of the helicity is violated to the maximum extent possible. Then, the BCFW recursion will be used to construct the 4 degree scattering amplitude from a 3 point amplitudes which we can easily compute using the *spinor helicity formalism.*

2. Spinor helicity formalism and color decomposition

In the usual Feynman approach in quantum field theory, there is too much redundancy among momenta and polarizations in the calculation of scattering amplitudes using Lorentz 4-vectors, especially when one has to deal with particles with spin. That arises the need to find new variables to represent the kinematics. It appears that one can organize efficiently the spin quantum numbers of external states in the *helicity basis.* In high energy process, almost all fermions are ultra-relativistic, behaving as if they were massless. Massless fermions that interact through gauge interactions have a conserved helicity, which we can exploit by computing in the helicity basis [3]. Although vector particles like photons and gluons do not have conserved helicity, it turns out that the most helicity-violating processes one can imagine are zero at tree level. That is, vectors can be written as *bispinors* $p_{a\dot{b}} = p_\mu \sigma^\mu_{a\dot{b}}$ where $\sigma^\mu = (\sigma^0, \sigma^j)_{j=1,2,3}$ are the usual Pauli matrices. For massless particles, $\det(p_{a\dot{b}}) = 0$, so we can write the momentum as a product of *Weyl spinors* [4] $p_{a\dot{b}} = \lambda_a \tilde{\lambda}_{\dot{b}}$. We can define the angle and square product as:

$$\langle ij \rangle = \epsilon_{ab} \lambda_i^a \lambda_j^b, \qquad [ij] = \epsilon_{\dot{a}\dot{b}} \tilde{\lambda}_i^{\dot{a}} \tilde{\lambda}_j^{\dot{b}}. \tag{1}$$

As a result, the usual Lorentz invariant product is written as $s_{ij} = \langle ij \rangle [ij]$. And for vector particles which have polarizations, the polarizations can be expressed as:

$$\epsilon_k^-(p_i) = -\frac{1}{\sqrt{2}} \frac{[k\gamma^\mu i\rangle}{[ki]}, \qquad \epsilon_k^+(p_i) = -\frac{1}{\sqrt{2}} \frac{[i\gamma^\mu k\rangle}{\langle ki \rangle}. \tag{2}$$

These equations satisfy $\epsilon_k^+(p_i)\epsilon_k^-(p_i) = -1$ and $p_i \epsilon_k^\pm(p_i) = 0$, where k is the *reference momentum.* In order to deal with the color factors in QCD, one can do a factorization of the scattering amplitude. This allows us to compute separately the *kinematic parts* and the *color-structures* of the scattering amplitude. For a process involving $(n-2)$ gluons and 2 fermions, one can do the following decomposition:

$$\mathcal{A}_n(1_q, 2_g, ..., n_{\bar{q}}) = g_s^{n-2} \sum_{\sigma \in S_n} \left(T^{\sigma(2)} T^{\sigma(3)} ... T^{\sigma(n-1)} \right) A_4[1, \sigma(2), ..., \sigma(n-1), n], \tag{3}$$

where the sum is over the permutations of the $(n-2)$ external legs. The color-independent amplitude $A_4[1, ..., n]$ is called ***color-ordered amplitude*** which is gauge invariant.

3. Computing the cross section using BCFW recursion

Let us now compute the cross section of the $qg \to qg$ process. Let us consider all the momentum to be outgoing. Thus, instead we compute $qgg\bar{q}$ and apply the crossing symmetry at the end. Denoted by 1 and 4 are the quark and the anti-quark, and 2 and 3 the two gluons with color indices a and b respectively. By virtue of the color decomposition formula, we can write down the full expression of the scattering amplitude:

$$\mathcal{A}_4(1_q, 2_g, 3_g, 4_{\bar{q}}) = g_s^2 T^a T^b A_4[1_q, 2_g, 3_g, 4_{\bar{q}}] + g_s^2 T^b T^a A_4[1_q, 3_g, 2_g, 4_{\bar{q}}]. \tag{4}$$

IOP Conf. Series: Journal of Physics: Conf. Series **889** (2017) 012021 doi:10.1088/1742-6596/889/1/012021

Let us make a choice of helicity such that we have a *Maximally Helicity Violating* (MHV)-amplitude, which means that we have exactly two particles with negative helicity, say $(-,-,+,+)$. Since different helicity do not interfere, as shown by the color decomposition of the full amplitude, to get the full answer of the amplitude one only has to compute the squares of all the possible helicity amplitude which can contribute to the process. The idea of *BCFW construction* is based on a linear momentum shift in complexified momentum space [5]. Let us consider the $[2,3\rangle$-shift:

$$|\widehat{2}] = |2] + z|3],\ |\widehat{2}\rangle = |2\rangle \quad \text{and} \quad |\widehat{3}\rangle = |3\rangle - z|2\rangle,\ |\widehat{3}] = |3]. \tag{5}$$

Hence, we have the following **BCFW on-shell diagrams**:

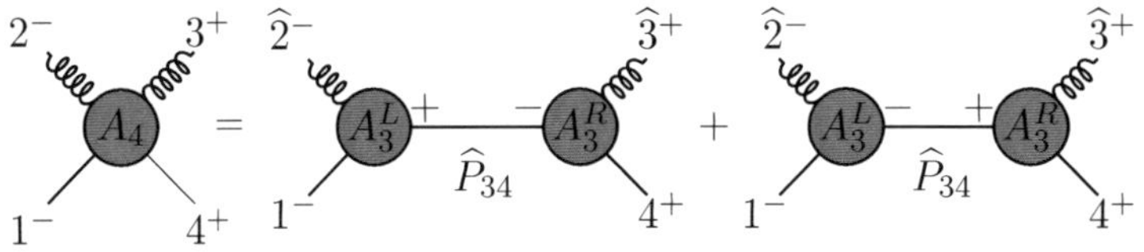

Figure 1. BCFW diagram for the $qgg\bar{q}$ process.

which can be mathematically expressed as:

$$A_4[1_q^-, 2_g^-, 3_g^+, 4_{\bar{q}}^+] = A_L^3[1_q^-, \widehat{2}_g^-, \widehat{P}_{34}^+]\frac{1}{P_{34}^2}A_R^3[-\widehat{P}_{34}^-, \widehat{3}_g^+, 4_{\bar{q}}^+]. \tag{6}$$

Here, the internal momentum $\widehat{P}_{34}$ is on-shell. The second diagram on the right hand-side vanishes and does not contribute to the calculation. Indeed, amplitudes with all the particles having the same helicity vanish $A_L^3[1_q^-, \widehat{2}_g^-, -\widehat{P}_{34}^-] = A_R^3[-\widehat{P}_{34}^+, \widehat{3}_g^+, 4_{\bar{q}}^+] = 0$. Using the momentum conservation to cancel the *momentum term* in the expression of the polarization, we end up with an expression of the *color-ordered subamplitude* in terms of the internal momentum $\widehat{P}_{34}$:

$$A_L^3[1_q^-, \widehat{2}_g^-, \widehat{P}_{34}^+] = \frac{\langle 12\rangle^2}{\langle \widehat{P}_{34}\rangle}, \quad \text{and} \quad A_R^3[-\widehat{P}_{34}^-, \widehat{3}_g^+, 4_{\bar{q}}^+] = \frac{[34]^2}{[4\widehat{P}_{34}]}. \tag{7}$$

Therefore, the total color-ordered amplitude can be written as:

$$A_4[1_q^-, 2_g^-, 3_g^+, 4_{\bar{q}}^+] = \frac{\langle 12\rangle^2}{\langle \widehat{P}_{34}\rangle}\frac{1}{\langle 34\rangle[34]}\frac{[34]^2}{[4\widehat{P}_{34}]}. \tag{8}$$

Since the internal momentum $\widehat{P}_{34}$ is *on-shell*, we have $\langle\hat{3}4\rangle[\hat{3}4] = 0$, and using the *shift equations* in (5), we find that the complex component z is equal to $\langle 34\rangle\langle 24\rangle^{-1}$. Hence,

$$|\hat{2}] = |2] + \frac{\langle 34\rangle}{\langle 24\rangle}|3],\ |\hat{2}\rangle = |2\rangle \quad \text{and} \quad |\hat{3}\rangle = |3\rangle - \frac{\langle 34\rangle}{\langle 24\rangle}|2\rangle,\ |\hat{3}] = |3]. \tag{9}$$

Furthermore, considering $\langle\hat{3}4\rangle[\hat{3}4] = 0$, since $[\hat{3}4]$ cannot be equal to zero, we have $|\widehat{3}\rangle$ is collinear to $|4\rangle$. Thus, we can write $|\hat{3}\rangle = k|4\rangle$. Using the new shift equations in (9), it follows that $k = \langle 23\rangle\langle 24\rangle^{-1}$. However, according to our definition, $\widehat{P}_{34} = k|4\rangle[3| + |4\rangle[4|$. We can therefore write the internal on-shell momentum only in terms of the momentum of the external particles. So, we have:

$$A_4[1_q^-, 2_g^-, 3_g^+, 4_{\bar{q}}^+] = \frac{\langle 12\rangle^3\langle 42\rangle}{\langle 12\rangle\langle 23\rangle\langle 34\rangle\langle 41\rangle} \quad \text{and} \quad A_4[1_q^-, 3_g^+, 2_g^-, 4_{\bar{q}}^+] = \frac{\langle 12\rangle^3\langle 42\rangle}{\langle 13\rangle\langle 32\rangle\langle 24\rangle\langle 41\rangle}. \tag{10}$$

IOP Conf. Series: Journal of Physics: Conf. Series **889** (2017) 012021 doi:10.1088/1742-6596/889/1/012021

To get the expression of $A_4[1_q^-, 3_g^+, 2_g^-, 4_{\bar{q}}^+]$, we have to swap 2 and 3 in the denominator from the $A_4[1_q^-, 2_g^-, 3_g^+, 4_{\bar{q}}^+]$-amplitude. Therefore, the expression of the full scattering amplitude can be written as:

$$\mathcal{A}_4(1_q, 2_g, 3_g, 4_{\bar{q}}) = g_s^2 \left[T^a T^b \frac{\langle 12\rangle^3 \langle 42\rangle}{\langle 12\rangle\langle 23\rangle\langle 34\rangle\langle 41\rangle} + T^b T^a \frac{\langle 12\rangle^3 \langle 42\rangle}{\langle 13\rangle\langle 32\rangle\langle 24\rangle\langle 41\rangle} \right]. \tag{11}$$

Squaring this amplitudes, taking into account the factors for averaging over the initial colors and helicities, adding the case with two gluons exchanged, and applying the crossing symmetry rules we get the final expression of the cross section of the process:

$$\frac{d\sigma}{d\cos\theta}(qg \to qg) = \frac{2}{9}\frac{\pi\alpha_s^2}{s}\left[-\left(\frac{s^2+u^2}{su}\right) + \frac{9}{4}\left(\frac{s^2+u^2}{t^2}\right)\right]. \tag{12}$$

4. Summary and Outlook

We have performed to derive correctly and in a much more efficient way the cross section of the $qg \to qg$ process [6]. The fact that the spinor helicity formalism fully exploit the properties of gauge invariance allows us to better organize the diagrammatic expansion. The BCFW on-shell recursion is a powerful tool to compute higher order scattering amplitude from a few number of external legs. The BCFW recursion relations can be used to build a generalization formula for scattering amplitude involving n external legs. For the case of $(n-2)$-gluons and 2-quarks, we have the generalization:

$$A_n[1_q^-, 2_g^+, ..., i_g^-, ..., n_{\bar{q}}^+] = \frac{\langle 1i\rangle^3 \langle ni\rangle}{\langle 12\rangle\langle 23\rangle ... \langle (n-1)n\rangle\langle n1\rangle}. \tag{13}$$

The aim of the project (AIMS essay) is to understand the energy loss of light quark passing through a Quark-Gluon Plasma. When a quark interacts with a thermal gluon, this leads to a emission of radiated gluons. The decrease of the energy of the quark is due to these emitted radiations. For that, we consider the following process:

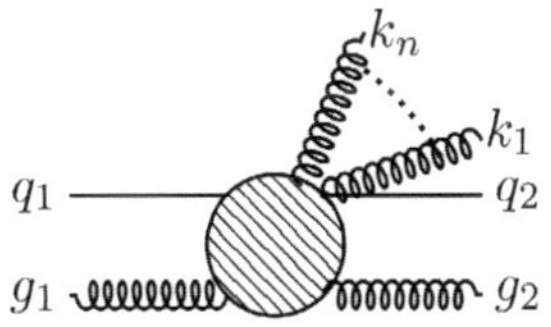

Figure 2. $q_1 g_1 \to q_2 g_2$ process with n bremsstrahlung gluons.

The idea is to start with 1 gluon radiation and then increase accordingly the number of bremsstrahlung gluons and we hope to generalize the concept to order n. We assume that the gluon radiation is *soft* and *collinear* to the outgoing quark. At *soft gluon limit* [7], we can factorize the amplitude into a divergent piece that depends on the energy and the angle of the emitted gluon (say k_s), and a second piece which is the amplitude omitting the soft gluon $\mathcal{A}_n(1, ..., i, k_s^{\pm}, j, ..., n) \to soft(i, k_s^{\pm}, j)\mathcal{A}_{n-1}(1, ..., i, j, ..., n)$. The expression of the *soft-amplitude* [8] is given by the MHV-formula. Similarly, if the gluon radiation is collinear to the outgoing quark, the amplitude factorizes into an amplitude with one fewer external state and a *splitting amplitude* depending on the particle going collinear and the internal state $\mathcal{A}_n(1, 2, \cdots, a, b, \cdots, n) \to \sum_{\lambda=\pm} split(a, b)\mathcal{A}_{n-1}(P^{\lambda}, 1, ..., n)$. These factorization properties of MHV-amplitudes will be used in the project.

IOP Conf. Series: Journal of Physics: Conf. Series **889** (2017) 012021 doi:10.1088/1742-6596/889/1/012021

Acknowledgments
The authors wish to thank the SA-CERN consortium and the South African National Research Foundation (NRF) for their generous support. We would also like to express our deepest appreciation to the African Institute for Mathematical Sciences (AIMS) and the University of Cape town for the opportunity to work on this project.

References
[1] M Gyulassy, P Levai, and I Vitev 2000 *Nucl. Phys. B* **571**
[2] R Keith Ellis, K Melnikov and G Zanderighi 2009 Phys. Rev. D **80** 094002
[3] Z Ber, L J Dixon and D A Kosower 2007 *Annals. Phys.* **322** 1587
[4] R. Britto, F. Cachazo and B. Feng, Nucle. Phys. B **715**, 499.
[5] Henriette Elvang and Yu-Tin Huang 2014 *arXiv:*1308.1697v2
[6] Hardik P Trivedi, Pallavi B, Anil Kumar, Lalit K Gupta, Jai Prakash Gupta, Krishna Chandra, Than Singh Saini and Archna Kansal 2012 *Nucl. Phys.* **57**
[7] M Mangano, S Parke and Z Xu 1988 *Nucl. Phys. B* **298** 653
[8] T G Birthwright, E W N Glover, V V Khoze and P Marquard 2005 *JHEP* **05**, **015**

IOP Conf. Series: Journal of Physics: Conf. Series **889** (2017) 012022 doi:10.1088/1742-6596/889/1/012022

Thermal Field Theory in Small Systems

W. A. Horowitz

Department of Physics, University of Cape Town, Private Bag X3, Rondebosch 7701, South Africa

E-mail: wa.horowitz@uct.ac.za

Abstract. We compute the finite size corrections to the partition function in a Cartesian space of finite extent in M directions and of infinite extent in $D - M$ directions for a massless, non-interacting scalar field theory. We then use this partition function to compute numerically the energy density, pressure, entropy density, and speed of sound for this theory for $M = 1$, 2, and 3 for $D = 3$ total spatial dimensions. The finite size corrections for the speed of sound are $\sim 600\%$, which indicates the need to consider these corrections in hydrodynamic simulations of small collision systems in high energy nuclear physics.

1. Introduction

The goal of heavy ion phenomenology is to understand the non-trivial, emergent, moderate-body physics of the strong nuclear force. This is an immodest goal as there isn't even a first principles calculation of the phase diagram of many of the most important many-body systems in QED, including water.

Approximately a microsecond after the Big Bang, the universe had cooled to a trillion degrees, or a few hundred MeV in natural units, a hundred thousand times hotter than the center of the sun [1]. We are precisely interested in these conditions, when, for instance, lattice simulations suggest a phase transition from normal nuclear matter to what has become known as the quark-gluon plasma (QGP) [2,3].

At asymptotically high temperatures, asymptotic freedom demands that the relevant degrees of freedom are (at most slightly thermally modified) quarks and gluons. But at these moderate temperatures of a few hundred MeV, the picture is not so clear. Lattice data shows a $\sim 15\%$ deviation from the Stefan-Boltzmann ideal gas limit of QCD for quantities such as the energy density and pressure of nuclear matter [2,3]. These slight but non-zero differences may be rather equally well attributed to the dynamics of nearly zero coupling [4] or of nearly infinite coupling [5].

Amazingly, we are able to experimentally probe these theoretical ideas at colliders such as the Relativistic Heavy Ion Collider (RHIC) at Brookhaven National Laboratory in Upton, New York, USA and at the Large Hadron Collider (LHC) in Geneva, Switzerland. The enormous wealth of data generated by these massive machines requires thousands of physicists to analyze, to put into a usable form for comparison to theoretical calculations.

A "standard model" of the collisions of large nuclei such as ^{197}Au and ^{208}Pb in heavy ion collisions is emerging in which the initial state physics of the color glass condensate (CGC) rapidly hydrodynamizes into a system for which nearly inviscid relativistic hydrodynamics with an equation of state given by lattice QCD provides an excellent description of the dynamics. The

IOP Conf. Series: Journal of Physics: Conf. Series **889** (2017) 012022 doi:10.1088/1742-6596/889/1/012022

system cools and hadronizes, with the final state nuclear matter rescattering until it is so dilute that it propagates freely until detection. This picture of the evolution of a heavy ion collision provides an excellent description of the distribution of the 99% of particles measured in heavy ion collisions, those particles with momenta in the direction transverse to the beam below ~few GeV/c [6]. Moreover, energy loss calculations—jet tomography, in which very high momentum particles created at the first instant of the collision that then subsequently propagate through the soft medium particles—appear to provide a very good qualitatively consistent picture of the distribution of particles above $\sim$ 15 GeV/c in transverse momentum [7].

A recent surprise is that this same "standard model" picture appears to also give a good description of the distribution of the low momentum particles in collisions of small systems such as a proton on a large nucleus [8]. This shocking conclusion raises two important questions: how are the thermodynamics changed in such a small system (previous calculations assumed a system in the thermodynamic limit) and how do particles lose energy in such a small system (previous calculations assumed a large system size)?

We will attempt to provide the start of an answer to the first question. The beginnings of an answer to the second question were discussed in [9].

2. Thermal Field Theory

Since relativistic hydrodynamics provides such a good description of the measured low momentum particles, and because the distribution of the low momentum particles show correlations amongst up to eight particles simultaneously [10], we conclude that the systems created in these heavy ion collisions behave collectively. We thus assume, presumably safely, that thermal quantum field theory [11, 12] provides a good description of the physics in these systems. In particular, we will use the technology of the canonical ensemble to compute various quantities such as the energy density, pressure, and speed of sound relevant for understanding the dynamics of quark-gluon plasma, even in systems far from the thermodynamic limit. Therefore we will concern ourselves with computing the partition function

$$Z = \sum_i e^{-E_i/T}, \tag{1}$$

where the sum over i is a sum over all possible states the system might take, E_i is the energy associated with the ith state, and T is the fixed temperature of the system.

The various quantities of interest may be computed from Z through

$$e = \frac{T^2}{V}\frac{\partial \ln Z}{\partial T} \tag{2}$$

$$p = T\frac{\partial \ln Z}{\partial V} \tag{3}$$

$$S = \ln Z + \frac{E}{T} \tag{4}$$

$$c_s^2 = \frac{\partial p}{\partial e}. \tag{5}$$

In order to make the connection between the usual partition function of statistical mechanics and quantum field theory, we exploit the relation

$$Z = \sum_i e^{-E_i/T} = \operatorname{tr} e^{-\beta \hat{H}} = \int \mathcal{D}\phi e^{-S_E} : \tag{6}$$

the sum over states is equal to the trace over the Hamiltonian operator weighted by $\beta \equiv 1/T$ which is, in turn, equal to the path integral over the Euclideanized action S_E. With the path integral in hand, one may use the methods of usual perturbative field theory, e.g. Feynman diagrams, even for systems for which $T \neq 0$.

3. Results for Systems of Infinite Size

For simple enough Lagrangians, one may analytically solve for the partition function Z and thus also for the various thermodynamic quantities of interest. For example, one may find E, p, etc. for a non-interacting scalar field

$$\mathcal{L} = \frac{1}{2}\partial_\mu\phi\partial^\mu\phi - \frac{1}{2}m^2\phi^2. \tag{7}$$

One method for finding Z for the above Lagrangian is to place the system in a box with sides of length L and expand $\phi(x)$ in the Fourier modes supported by the box. The calculation is simpler if we impose Neumann boundary conditions; then we may take

$$\phi(\tau,\vec{x}) = \sqrt{\frac{\beta}{V}}\sum_{n,\vec{p}}\tilde{\phi}(\omega_n,\vec{p})e^{i(\vec{p}\cdot\vec{x})+\omega_n\tau}. \tag{8}$$

For a massless theory in D spatial dimensions, one finds after the $T=0$ vacuum subtraction that

$$\ln Z(\beta,D,m) = -\frac{V_D\beta^{-D}}{(4\pi)^{\frac{D+1}{2}}}\sum_{\ell=1}^{\infty}(\frac{2m\beta}{\ell})^{\frac{D+1}{2}}K_{\frac{D+1}{2}}(m\beta\ell) \tag{9}$$

By expanding the modified Bessel function for small argument, $K_\nu(x) = 2^{\nu-1}\Gamma(\nu)x^{-\nu}+\mathcal{O}(x)$, one finds

$$\ln Z\big|_{\beta,D,m=0} = \frac{V_d}{\pi^{(D+1)/2}}\zeta(D+1)\Gamma(\frac{D+1}{2})\beta^{-D} \tag{10}$$

$$\rightarrow \frac{V_3\pi^2}{90}T^3,\ \text{for}\ D=3. \tag{11}$$

Therefore one immediately finds that

$$\begin{aligned}
p(D,T) &= \frac{\partial T\ln Z}{\partial V} = \frac{1}{\pi^{(D+1)/2}}\zeta(D+1)\Gamma(\frac{D+1}{2})T^{D+1} \rightarrow \frac{\pi^2}{90}T^4,\quad \text{for}\ D=3,\\
e(D,T) &= \frac{T^2}{V}\frac{\partial\ln Z}{\partial T} = d\frac{1}{\pi^{(D+1)/2}}\zeta(D+1)\Gamma(\frac{D+1}{2})T^{D+1} \rightarrow \frac{\pi^2}{30}T^4,\quad \text{for}\ D=3,\\
c_s(D,T) &= \sqrt{\frac{\partial p}{\partial e}} = \frac{1}{\sqrt{D}} \rightarrow \frac{1}{\sqrt{3}},\quad \text{for}\ D=3
\end{aligned} \tag{12}$$

4. Results for Systems of Finite Size

In order to model QGP-like systems of finite size, we choose to replace the Neumann boundary conditions with Dirichlet boundary conditions in M of the D spatial directions. The idea is that these Dirichlet boundary conditions model how the QGP created in heavy ion collisions is of a finite extent inside a larger vacuum.

In the above derivation of the partition function Eq. (9) we took

$$\sum_{\vec{p}} \rightarrow V_D\int\frac{d^Dp}{(2\pi)^D}. \tag{13}$$

With these M compactified directions we take

$$\sum_{\vec{p}} \to \sum_{\vec{\ell}=(\ldots,1,\ldots)}^{(\ldots,\infty,\ldots)} V_{D-M} \int \frac{d^{D-M}p}{(2\pi)^{D-M}}. \tag{14}$$

The logarithm of the partition function then becomes more complicated:

$$\begin{aligned} lnZ = V_{D-M}\Big(& \frac{\beta}{(2\pi)^{D+1}} \sum_{i=0}^{M} (-1)^i \frac{\pi^i}{i!(M-i)!} \sum_{\sigma} \left(L_{\sigma(i+1)} \cdots L_{\sigma(M)}\right) \Big\{ \\ & (1-\delta_{M,i}) \sum_{\vec{k}\in\mathbb{Z}^{M-i}\setminus\{0\}} \frac{1}{2}\Gamma\Big(\frac{D-i+1}{2}\Big)\Big[\frac{1}{\pi}\sum_{j=1}^{M-1} k_j^2 L_{\sigma(i+j)}^2\Big]^{-\frac{D-i+1}{2}} \Big\} \\ & + 2\frac{\beta^{-(D-M)}}{(2\pi)^{\frac{D-M+1}{2}}} \sum_{\vec{\ell}=(\ldots,1,\ldots)}^{(\ldots,\infty,\ldots)} \sum_{\ell=1}^{\infty} \Big(\frac{\tilde{\omega}_{\vec{\ell}}\beta}{\ell}\Big)^{\frac{D-M+1}{2}} K_{\frac{D-M+1}{2}}(\tilde{\omega}_{\vec{\ell}}\beta\ell)\Big), \end{aligned} \tag{15}$$

where

$$\tilde{\omega}_{\vec{\ell}} \equiv \Big(\sum_{i=1}^{M} \Big(\frac{\pi\ell_i}{L_i}\Big)^2\Big)^{1/2}. \tag{16}$$

In Eq. (15), the L_i are the finite extents of the space in each of the M compactified directions. The sum over σ is the sum over all permutations of the compactified directions, thus guaranteeing that Eq. (15) is symmetric under the interchange of any of the L_i.

One may compute analytically the result for, e.g., $D = 3$ and $M = 1$ from Eq. (15) to find that

$$\ln Z\Big|_{TL,D=3,M=1,m=0} = \frac{V_2}{L^2}\Big[\frac{\pi^2}{1440}\frac{1}{TL} + \frac{(TL)^2}{2\pi}\sum_{\ell=1}^{\infty}\Big\{\frac{\pi\ell}{TL}\mathrm{Li}_2(e^{-\frac{\pi\ell}{TL}}) + \mathrm{Li}_3(e^{-\frac{\pi\ell}{TL}})\Big\}\Big], \tag{17}$$

where $\mathrm{Li}_s(z)$ is the usual polylogarithmic function. One can show that Eq. (17) has the correct $TL \to \infty$ behavior:

$$\ln Z\Big|_{TL,D=3,M=1,m=0} \to \frac{V_2}{L^2}\frac{\pi^2(TL)^3}{90} = V_3\frac{\pi^2 T^3}{90}. \tag{18}$$

One may see the numerical results for various thermodynamic relations given the above partition function Eq. (15) in Fig. (1).

5. Conclusions and Outlook

As one can see from Fig. (1), the small system size corrections to the usual thermodynamic properties of an ideal gas can be very large. In particular, while the energy density, pressure, and entropy density differences are relatively small, $\sim 20\%$, the speed of sound increases by up to a factor of 6! It is difficult to understand 1) why the speed of sound is so much more sensitive to the small size corrections than the other thermodynamic quantities, 2) how to interpret a speed of sound that is larger than the speed of light in vacuum, and 3) at what length scale the derivation becomes invalid. These issues are currently under investigation. We are also examining the effect of a non-zero mass for the field.

Future work includes improving the calculation of the finite system size corrections to energy loss. Previous work [9] found the finite distance corrections to the usual DGLV energy loss formalism while keeping the infinite size GW model for the medium scattering centers. It will be very interesting to compute the finite size corrections to, e.g., the Debye screening mass.

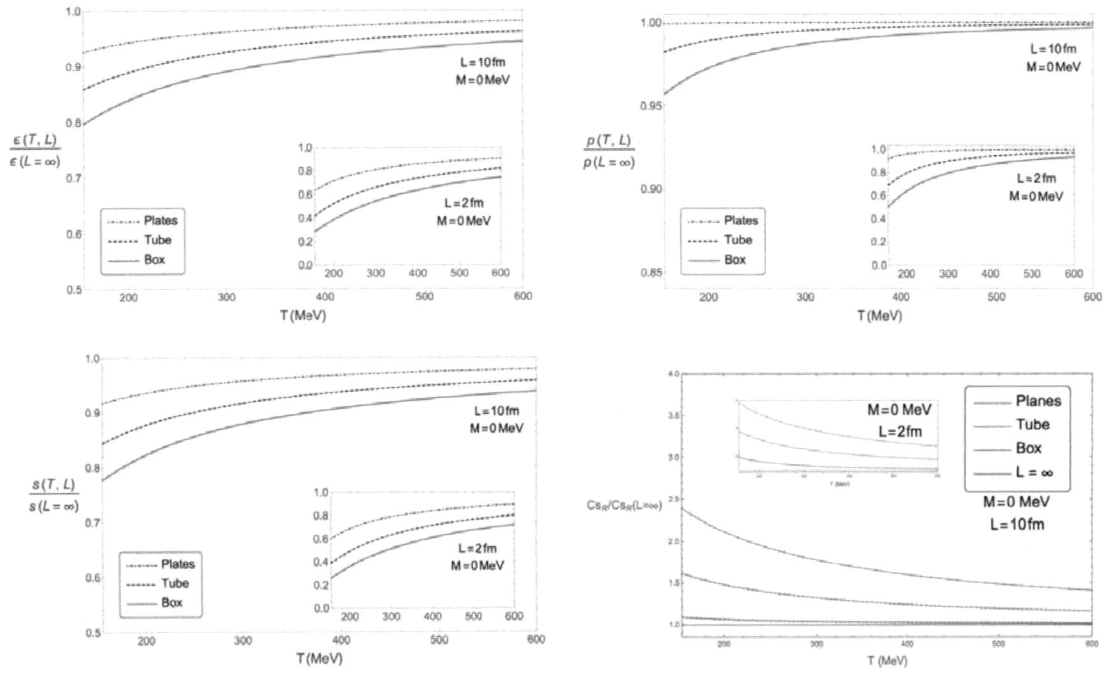

Figure 1: (Color online) Ratios of thermodynamic quantities in a finite size system to that of an infinite size system as a function of temperature for a massless, non-interacting scalar field theory. The large figures have a finite length of $L = 10$ fm while the insets have $L = 2$ fm. The three curves curves are for $M = 1$ (plates), $M = 2$ (tube), and $M = 3$ (box) compact dimensions in a $D = 3$ spatial dimensions universe. Clockwise from top left are: energy density, pressure, speed of sound, and entropy density.

6. Acknowledgments

The author wishes to thank the National Research Foundation and the SA-CERN collaboration for generous support.

References

[1] "The Frontiers of Nuclear Science, A Long Range Plan," arXiv:0809.3137.
[2] A. Bazavov *et al.*, *Phys. Rev.*, vol. D80, p. 014504, 2009.
[3] S. Borsanyi *et al.*, *JHEP*, vol. 11, p. 077, 2010.
[4] J. O. Andersen, L. E. Leganger, M. Strickland, and N. Su, *JHEP*, vol. 08, p. 053, 2011.
[5] S. S. Gubser, *Nucl. Phys.*, vol. A830, pp. 657C–664C, 2009.
[6] C. Gale *et al.*, *Phys. Rev. Lett.*, vol. 110, no. 1, p. 012302, 2013.
[7] W. A. Horowitz, *Nucl. Phys.*, vol. A904-905, pp. 186c–193c, 2013.
[8] R. D. Weller and P. Romatschke, arXiv:1701.07145.
[9] I. Kolbe and W. A. Horowitz, arXiv:1511.09313.
[10] V. Khachatryan *et al.*, *Phys. Rev. Lett.*, vol. 115, no. 1, p. 012301, 2015.
[11] J. I. Kapusta and C. Gale, *Finite-temperature field theory: Principles and applications.* Cambridge University Press, 2011.
[12] M. Laine and A. Vuorinen, *Basics of Thermal Field Theory.* Springer, 2016.

IOP Conf. Series: Journal of Physics: Conf. Series **889** (2017) 012023 doi:10.1088/1742-6596/889/1/012023

Physics with four-leptons at the LHC

Yesenia Hernández Jiménez

School of Physics, University of the Witwatersrand, Johannesburg 2050, South Africa

E-mail: yesenia@cern.ch

Abstract. Several analyses with four-leptons in the final state performed by ATLAS and CMS experiments at the LHC are summarized. These results include Standard Model measurements on ZZ and Higgs boson production, as well as beyond Standard Model searches. The four-lepton signature of these processes leads to a very clean final state with low background. These features enhance the interest of these analyses, which may be performed in the context of beyond the Standard Model searches for the Run 2 period. In particular, the resonant pair production of new heavy scalars, the Madala boson, decaying to four W-bosons will be discussed.

1. Introduction: Four-lepton searches at the LHC

Searches relying on four leptons in the final state use a very clean signature with low background contamination. Due to these features, different four-leptons analyses are developed in ATLAS and CMS experiments for such Standard Model (SM) measurements, as ZZ production and Higgs boson production as well as for exploring a wide range of new physics. ‡

2. Standard Model Analyses

Inclusive ZZ production cross section in proton-proton collisions at 8 TeV is measured by both CMS [1] and ATLAS [2] experiments. The former uses an integrated luminosity of 19.6 fb^{-1} and the measurements are performed in the leptonic decay modes ZZ to $lll'l'$, where $l = e, \mu$ and $l' = e, \mu, \tau$. The measured total cross section is $\sigma(pp \to ZZ) = 7.7 \pm 0.5(\text{stat.})^{+0.5}_{-0.4}(\text{syst.}) \pm 0.4(\text{theo.}) \pm 0.2(\text{lum.})$ pb for both Z bosons produced in the mass range $60 < m_Z < 120$ GeV. For ATLAS, the total cross section for ZZ events produced with both Z bosons in the mass range 66 to 116 GeV is measured from the combination of the two channels to be $7.3 \pm 0.4(\text{stat}) \pm 0.3(\text{syst}) \pm 0.2(\text{lumi})$ pb. Both results are consistent with the Standard Model prediction of $6.6^{+0.7}_{-0.6}$ pb.

‡ We don't consider here pair production of J/ψ or Υ states as well as such states as η, η', ρ, ω, ϕ which can also result to four-lepton final states.

IOP Conf. Series: Journal of Physics: Conf. Series **889** (2017) 012023 doi:10.1088/1742-6596/889/1/012023

2.1. Standard Model: Four-leptons Results in ATLAS

The ATLAS Collaboration measured four-lepton production cross section in the mass range from 80 to 1000 GeV [3] based on 20.3 fb^{-1} of pp collisions at $\sqrt{s} = 8$ TeV. The four-lepton events are produced in the decays of resonant Z and Higgs bosons and the non-resonant ZZ continuum originating from $q\bar{q}$, gg, and qg initial states. A total of 476 signal candidate events are observed with a background expectation of 26.2 ± 3.6 events, enabling the measurement of the integrated cross section and the differential cross section as a function of the invariant mass and transverse momentum of the four-lepton system. In the mass range above on-shell Z-boson pair production (180 GeV) the signal strength of the gluon-fusion component relative to its leading-order prediction is determined to be $\mu_{gg} = 2.4 \pm 1.0(\text{stat.}) \pm 0.5(\text{syst.}) \pm 0.8(\text{theory})$. Figure 1 shows the four-lepton invariant-mass distribution ($m_{4\ell}$) and the measured differential cross-section as a function of $m_{4\ell}$.

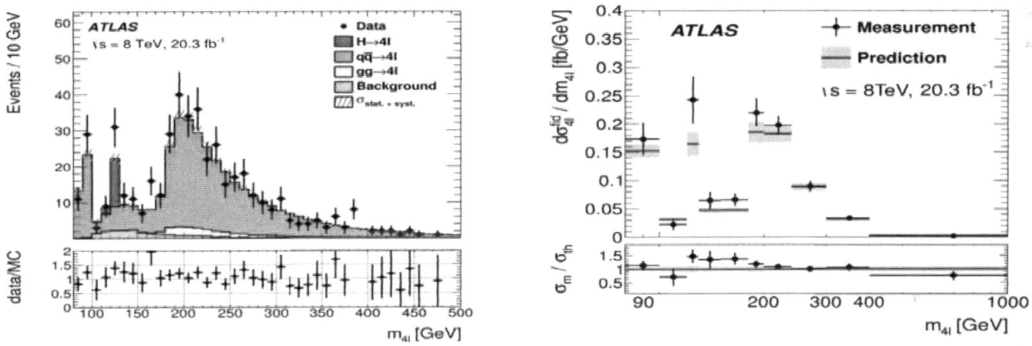

Figure 1: Four-lepton invariant-mass distribution (left) and differential cross-section as a function of $m_{4\ell}$ (right) measured by ATLAS [3]. Bottom plots show data/MC ratio.

2.2. $H \to ZZ^* \to 4\ell$ observation in the CMS and ATLAS experiments

The ATLAS Run 1 measurements of Higgs boson production rates and couplings in the decay channel $H \to ZZ^* \to 4\ell$ are described in Ref. [4]. These measurements were performed using pp collision data corresponding to integrated luminosities of 4.5 and 20.3 fb^{-1} at center-of-mass energies of 7 and 8 TeV, respectively. The $H \to ZZ^* \to 4\ell$ signal is observed with a significance of 8.1 standard deviations (σ), with an expectation of 6.2 σ, at $m_H = 125.36$ GeV, the combined ATLAS measurement of the Higgs boson mass from the $H \to \gamma\gamma$ and $H \to ZZ^* \to 4\ell$ channels. The production rate relative to the Standard Model expectation, the signal strength, is measured in four different production categories in the $H \to ZZ^* \to 4\ell$ channel. The measured signal strength, at this mass, and with all categories combined, is $1.44^{+0.40}_{-0.33}$.

The CMS Collaboration also observed at 7–8 TeV [5] the new boson as a narrow resonance with a local significance of 6.8 σ, a measured mass of $125.6 \pm 0.4(\text{stat}) \pm 0.2(\text{syst})$ GeV, and a total width less than 3.4 GeV at the 95% confidence level.

Table 1 quotes the number of events expected in the 118-129 GeV mass range and observed for the $m_H = 125$ GeV hypothesis for the four-lepton final states in ATLAS at $\sqrt{s} = 13$ TeV. Figure 2 summarizes the $H \to ZZ^* \to 4\ell$ Run-1 results for the observed local p_0 value obtained by ATLAS and CMS [6, 7].

Final State	Signal full mass range	Signal	ZZ^*	Z + jets, $t\bar{t}$ ttV,VVV, WZ	S/B	Expected	Observed
4μ	8.8 ± 0.6	8.2 ± 0.6	3.11 ± 0.30	0.31 ± 0.04	2.4	11.6 ± 0.7	16
$2e2\mu$	6.1 ± 0.4	5.5 ± 0.4	2.19 ± 0.21	0.30 ± 0.04	2.2	8.0 ± 0.4	12
$2\mu 2e$	4.8 ± 0.4	4.4 ± 0.4	1.39 ± 0.16	0.47 ± 0.05	2.3	6.2 ± 0.4	10
$4e$	4.8 ± 0.5	4.2 ± 0.4	1.46 ± 0.18	0.46 ± 0.05	2.2	6.1 ± 0.4	6
Total	24.5 ± 1.8	22.3 ± 1.6	8.2 ± 0.8	1.54 ± 0.18	2.3	32.0 ± 1.8	44

Table 1: Number of expected and observed candidates in the 118–129 GeV mass range for the $m_H = 125$ GeV hypothesis using four-lepton events in ATLAS [6].

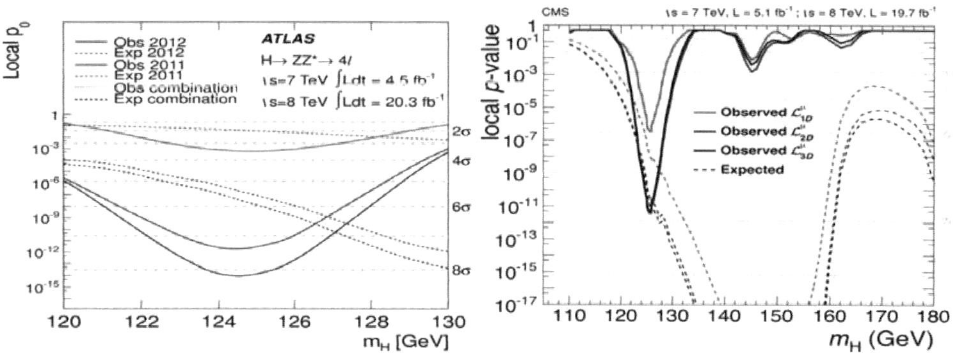

Figure 2: The observed local p_0-value for the combination of the 2011 and 2012 data sets as a function of m_H for ATLAS (left) and CMS (right) [4, 5].

2.3. $ZH \to ZWW^* \to 4\ell + 2\nu$ *Results in ATLAS*

A search for Higgs boson production in association with a Z boson, in the $H \to WW^*$ decay channel, is performed with a data sample collected with the ATLAS detector at the LHC in proton-proton collisions at centre-of-mass energies $\sqrt{s} = 7$ TeV and 8 TeV, corresponding to integrated luminosities of 4.5 fb^{-1} and 20.3 fb^{-1}, respectively [8].

The strategy of the analysis is to select four-lepton events with total charge zero, which are categorized according to the number of same-flavour opposite signed (SFOS) lepton pairs: 1-SFOS ($3e + 1\mu$, $1e + 3\mu$) and 2-SFOS ($4e$,4μ,$2e + 2\mu$). The analysis exploits the kinematics of the opposite-sign lepton pair from the Higgs boson decay candidate by requiring low angular separation between the leptons in the Z-boosted frame. The Run-I results for the signal significance (Z_0) and the $H \to WW^*$ signal strength (μ) evaluated at the end of the event selection are quoted in Table 2.

Category	Signal significance Z_0: Exp. Z_0	Obs. Z_0	Obs. Z_0	Observed signal strength μ: μ	Tot. err. +	Tot. err. −	Syst. err. +	Syst. err. −	μ
ZH only	0.30	2.0		**5.1**	4.3	3.1	1.9	0.89	
			0 1 2 3 4 5 6 7						0 1 2 3 4 5 6 7 8 9 10

Table 2: The signal significance and the $H \to WW^*$ signal strength evaluated in the signal region, combining the 8 TeV and 7 TeV data [8].

2.4. $t\bar{t}H$ results with four-leptons in ATLAS

The Yukawa coupling of the Higgs boson to the top quark is a key parameter of the Standard Model, and can be constrained using the associated production process $pp \to t\bar{t}H + X$. A search for this process using final states with multiple leptons has been performed in ATLAS using Run 1 and Run 2 data samples [9, 10]. The analysis of the four-lepton final state is statistically limited and only one event is observed with Run 1 data. This leads to a measured signal strength of $1.8^{+6.9}_{-2.0}(\text{tot.})^{+6.8}_{-2.0}(\text{stat.})$.

3. Beyond Standard Model Searches

3.1. High Mass $S \to ZZ^* \to 4\ell$ with Run 2 data

A search for scalar resonances decaying into the four-lepton final state is performed in ATLAS [6] and CMS [11] using Run 2 data at 13 TeV. Related $S \to ZZ$ analyses closely follow the SM Higgs strategy but extending the search to the higher mass range 200−800 GeV to cover different S mass hypotheses. Figure 3 shows the distribution of the four-lepton reconstructed invariant mass in the high-mass range for both ATLAS and CMS experiments and the ATLAS 95% confidence level limits on the S boson production of the additional heavy Higgs-like boson assuming the Narrow Width Approximation (NWA) [12] for different mass hypotheses in the range 200 − 1000 GeV.

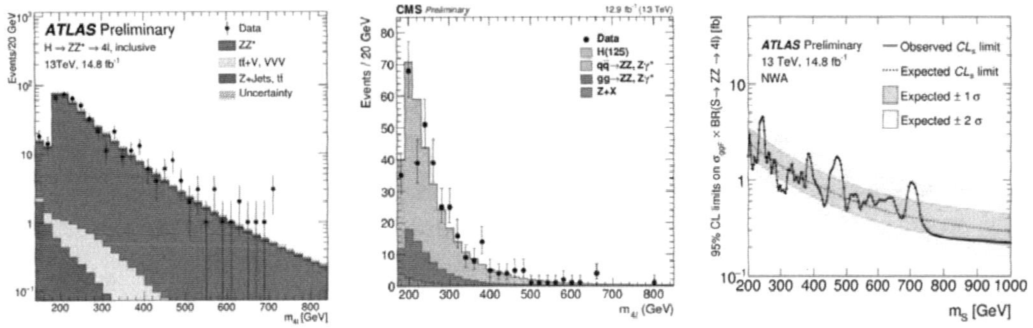

Figure 3: Invariant mass of the four-lepton system for high mass $S \to ZZ^* \to 4\ell$ searches at ATLAS (left) and CMS (middle) [6, 11]. The 95% confidence limits as function of m_S for an additional narrow heavy scalar in ATLAS (right).

IOP Conf. Series: Journal of Physics: Conf. Series **889** (2017) 012023 doi:10.1088/1742-6596/889/1/012023

3.2. The Madala Boson Hypothesis

Higgs boson searches at the LHC can be extended assuming the validity of the so-called Madala boson model [13–16]. This model introduces an additional doublet of two hypothetical scalars H and S with $2m_h < m_H < 2m_t$ and $m_h < m_S < m_H$, where m_h and m_t represent the masses of the Higgs boson and top-quark, respectively. In addition, the introduction of new HSS vertex leads to the $H \to SS$ decay with S being a portal to dark matter (DM) through its interaction with χ, a DM candidate.

This feature motivates a search in ATLAS for the $H \to SS \to 4W$ decay. Specially the presence of four leptons in the final state leads to a unique final state with a very clean signature due to low expected background contribution. The strategy of the search relies on the kinematics of the four-lepton system as well as on the kinematics of the lepton pairs. Figure 4 compares different kinematic quantities from simulation of $H \to SS, Sh$ process with four-lepton final states.

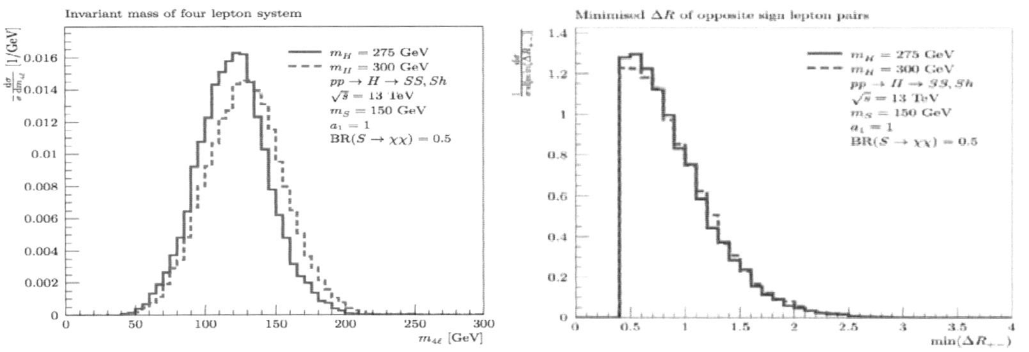

Figure 4: The invariant mass of the four lepton system (left) and the minimum distance of leptons (right) for the signal $pp \to H \to SS, Sh$ processes from MC simulation for $m_H = 275(300)$ GeV in blue(red) assuming $\mathrm{BR}(S \to \chi\chi) = 0.5$ [14].

4. Conclusions

Results of some four-lepton analyses performed by the ATLAS and CMS experiments at the LHC are presented. They include $Z^*Z^* \to 4\ell$ and $H \to ZZ^* \to 4\ell$ SM production cross sections as well as rates of associated ZH and ttH-production with four-lepton final states. Apart of the SM analyses, new physics can be also explored using events with four leptons. Upper limits on the high mass $S \to ZZ$ production cross section in the NWA model are quoted and for new massive scalar searches, as the Madala boson, are discussed. The 2017 Run 2 data would become crucial to verify the obtained results so far, to measure other Higgs production mechanisms (ZH and $t\bar{t}H$) and to provide enough statistics to obtain for the first time results on new physics as the Madala boson.

IOP Conf. Series: Journal of Physics: Conf. Series **889** (2017) 012023 doi:10.1088/1742-6596/889/1/012023

References

[1] CMS Collaboration. *Phys. Lett.*, B740:250–272, 2015.

[2] ATLAS Collaboration. *Journal of High Energy Physics*, 01:099, 2017.

[3] ATLAS Collaboration. *Phys. Lett.*, B753:552–572, 2016.

[4] ATLAS Collaboration. *Phys. Rev. D*, 91:012006, Jan 2015.

[5] CMS Collaboration. *Journal of High Energy Physics*, 04:005, 2016.

[6] ATLAS Collaboration. Technical Report ATLAS-CONF-2016-079, CERN, Geneva, Aug 2016.

[7] CMS Collaboration. Technical Report CMS-PAS-HIG-16-033, CERN, Geneva, 2016.

[8] ATLAS Collaboration. *Journal of High Energy Physics*, 2015(8):137, 2015. ISSN 1029-8479.

[9] ATLAS Collaboration. *Physics Letters B*, 749:519 – 541, 2015. ISSN 0370-2693.

[10] ATLAS Collaboration. Technical Report ATLAS-CONF-2016-058, CERN, Geneva, Aug 2016.

[11] CMS Collaboration. Technical Report CMS-PAS-HIG-16-033, CERN, Geneva, 2016.

[12] D. Berdine, N. Kauer, and D. Rainwater. *Phys. Rev. Lett.*, 99:111601, 2007.

[13] S. von Buddenbrock et al. 2015. URL `https://arxiv.org/abs/1506.00612`.

[14] S. von Buddenbrock et al. *Eur. Phys. J.*, C76(10):580, 2016.

[15] M. Kumar et al. In *Workshop on High Energy Particle Physics (HEPPW2016) Johannesburg, South Africa, February 8-10, 2016*, 2016. URL `https://inspirehep.net/record/1425948/files/arXiv:1603.01208.pdf`.

[16] M. Kumar et al. In *61st Annual Conference of the South African Institute of Physics (SAIP2016) Johannesburg, South Africa, July 4-8, 2016*, 2016. URL `https://inspirehep.net/record/1480869/files/arXiv:1608.03466.pdf`.